Robert F. Schleif and Pieter C. Wensink

Practical Methods in Molecular Biology

With 49 Figures

Springer-Verlag
New York Heidelberg Berlin

Robert F. Schleif
Department of Biochemistry
Brandeis University
Waltham, Massachusetts 02254
U.S.A.

Pieter C. Wensink
Rosenstiel Center and
Department of Biochemistry
Brandeis University
Waltham, Massachusetts 02254
U.S.A.

Sponsoring Editor: Philip Manor
Production: Kate Ormston

Library of Congress Cataloging in Publication Data
Schleif, Robert F.
Practical methods in molecular biology.
Bibliography: p.
Includes index.
1. Molecular biology—Technique. I. Wensink, Pieter C. II. Title. [DNLM: 1. Molecular biology—Methods. QH 506 S339p]
QH506.S34 574.8′8′028 81-9325
AACR2

Printed in the United States of America.

9 8 7 6 5 (Fifth Printing, 1987)

ISBN 0-387-90603-7 Springer-Verlag New York Heidelberg Berlin
ISBN 0-540-90603-7 Springer-Verlag Berlin Heidelberg New York

To our parents

Preface

This volume has evolved from a laboratory methods book that one of us first compiled nearly fifteen years ago. Since that time the book has undergone many minor revisions in order to include new methods and updated versions of older methods. The result has been an increasingly useful and more widely circulated book. However, the recent series of technological explosions generally lumped together under the name of "recombinant DNA technology" has been a turning point in the evolution of this previously underground publication. Minor revisions will no longer do. To keep the book useful we have had to make major revisions and additions. The result is a dramatically expanded book that should be more useful to more people. The larger size and wider usefulness of the book have made this more formal publication seem a reasonable step to take.

One of the reasons that this volume should be useful to many people is that it includes only procedures that have been used repeatedly by us and that have proven highly reliable both to ourselves and to others in our laboratories. We have made this decision to include only highly reliable recipes that we ourselves use because we want the book to be a dependable resource to others. While this dependability is a major virtue of the book, it does force us to make a few sacrifices. We have had to exclude some important procedures. Also, as is frequently the case in experimental work, the most dependable procedures usually lack the sophistication or efficiency of refined, state-of-the-art methods.

This book covers a broad range of techniques. The recipes range from how to grow bacteria to techniques for clone selection by hybridization and for in vitro translation of messenger RNAs. Furthermore, we have presented them in a greater depth than is normally found in contemporary scientific publications. As a result, the procedures should be useful to someone with no prior experience with the techniques.

It is worth noting that the book contains a large number of recipes that are relevant to genetic engineering. Techniques are described for such routine manipulations as DNA purification, concentration, and quantitation as well as for some of the procedures necessary for cloning genes from higher organisms. However, this book does not include the very extensive collection of techniques for DNA sequencing which are described in *Methods in Enzymology,* Vol. 65, Part I, edited by Grossman and Moldave, 1980. It also does not contain the vast collection of detailed genetic techniques which are contained in Jeffrey Miller's book, *Experiments in Molecular Genetics.* The reader should use these sources for DNA sequencing or for extensive genetic analyses.

The physical design of the book is intended to facilitate its use in the laboratory. The wide margins and hard but not glossy paper should allow users to pencil in and then to change their own modifications of the recipes. The clean and somewhat oversize type will ease reading by the harassed experimenter.

This book could not have been written without the information provided by a large number of people in the field. We are fortunate to be working in a field that has a tradition of free interchange of even the most intimate details of current experimental methods. We thank all of those who have given us help in developing these recipes.

Robert F. Schleif
Pieter C. Wensink

Contents

Chapter 1 Using *E. coli* 1
Strains 1
Strain Purity 1
Sources of Strains 2
Commonly Used Strains 2
Pedigrees and Genetic Maps 2
Strains for Physiological Measurements 2
Storage of Strains 3
Cell Growth 3
Properties of Bacteria 3
Phage Contamination 5
Use of Cells for Physiological Measurements 6
Use of Cells for Genetic Purposes 7
Growing Cells for the Purification of Molecules 9
Measuring Cell Density 9
Growing Large Quantities of Cells 11
Opening Cells 15
Sonicating 15
Grinding in Alumina 15
Grinding with Glass Beads 15
Opening Cells with Lysozyme 16
Radiolabelling Cells 16
Nitrosoguanidine Mutagenesis 17
Penicillin Selection 18
Curing Cells of F-Factors 19
Curing with Acridine Orange 19
Selecting Spontaneously Cured Cells 20
Curing with Sodium Dodecyl Sulfate 20
Curing with High Temperature 20
Making Cells Streptomycin-Resistant 20
Crossing *rec*A into Cells 21
Phage P1 Transduction of Genetic Markers 22
Large-Scale Genetic Crosses 24
Using Transposons in Strain Construction 26

Chapter 2 Bacteriophage Lambda **27**
Two Useful Mutants 27
CI_{857} 27
S_7 28
Titering 28
Growing Plate Stocks 29
Large-Scale Growth in Liquid 31
Purification 33
Genetic Crosses 35
Scoring Plaques 36
Making Strains Lambda-Resistant 37
Testing Colonies for the Ability to Grow Lambda 38
Making Lysogens 38
Testing Lysogen Candidates 40
Streaking for Single Plaques 40
Selecting Deletions 41

Chapter 3 Enzyme Assays **43**
β-Galactosidase 43
RNA Polymerase 45
Arabinose Isomerase 46
Lysozyme 50
Ribulokinase 52
E. coli-Coupled Transcription-Translation System 56

Chapter 4 Working with Proteins **61**
Ammonium Sulfate Precipitation of Proteins 62
Removing Nucleic Acids by Phase Partition 64
Bulk Separation of Proteins from Nucleic Acids 65
Purification of Protein-Bound DNA or RNA 67
Columns, Fraction Collectors, and Plumbing 68
Ion Exchange Chromatography and Gel Filtration 69
Preparing Ion Exchange Resins 69
Pouring Columns 70
Loading and Eluting Columns 71
Determining Protein Concentration 74
Optical Density 74
Biuret 74
Lowry 75
Lowry for Dilute Samples 75
Concentrating Protein Solutions 76
Stabilizing Proteins 77
Polyacrylamide Gel Electrophoresis of Proteins 78
Making a Gel Sandwich 80
SDS-10% Acrylamide Gel with Stacking Gel 81
Urea-SDS-Acrylamide Gradient Gels 84
Staining Gels 87
Fluorography of $[^3H]$- or $[^{35}S]$-Labelled Proteins in Acrylamide Gels 87
Recovering PPO from Solution in DMSO 88

Chapter 5 Working with Nucleic Acids **89**
Measuring Nucleic Acid Concentration and Purity 89
Optical Methods 89
Fluorescence Method 90
Storing DNA 92
Cleaning DNA 93
Cleaning by DEAE 94
Cleaning by Hydroxyapatite 94
Precipitating DNA with Ethanol 95
TCA Precipitation Assay 96
Precipitation and Size Fractionation of DNA with Polyethylene Glycol 97
Isolating *E. coli* DNA 98
Isolating Lambda DNA 99
Phenol Extraction 99
SDS Extraction 100
Isolating Plasmid DNA 101
Large-Scale Plasmid Isolation 106
Isolating *Drosophila* DNA 108
Preparing Nucleoside Triphosphate Solutions 111
Chromatographic Analysis of Nucleosides 112
Gel Electrophoresis of DNA 114
Agarose Gel Electrophoresis 115
Polyacrylamide Gel Electrophoresis 120
Staining and Photographing Gels 121
Extracting DNA from Acrylamide and Agarose Gels . 122
Mapping Restriction Endonuclease Sites on DNA 125

Chapter 6 Constructing and Analyzing Recombinant DNA **129**
Joining the Ends of DNA Molecules 129
Making Blunt Ends by S1 Digestion 130
Ligation with T4 DNA Ligase 130
Lambda Exonuclease Digestion to Yield Free 3′ Ends . 132
Addition of Homopolymer Tails 133
E. coli Transformation with Plasmid DNA 134
Storing Strains that Contain Plasmids 136
Cycloserine Selection of Recombinant Plasmids 136
In Vitro Radiolabelling of DNA and RNA 137
Radiolabelling DNA by Nick Translation 138
Synthesizing Radiolabelled RNA Complementary to DNA 140
Synthesizing Radiolabelled DNA Complementary to RNA 141
Radiolabelling the 5′ Ends of RNA or DNA 143
General Aspects of Nucleic Acid Hybridization Reactions 145
Screening Recombinant DNA Clones by Nucleic Acid Hybridization 146
Screening Phage Plaques 147
Screening Colonies 149

Isolating DNA from a Single Colony 151
Southern Transfers 152
Selecting RNA Complementary to a DNA 156
Preparing DNA 157
Preparing NBM Paper 157
Converting NBM Paper to DBM Paper and Binding the DNA 158
Hybridizing RNA to DBM-Bound DNA 158
Recovering the RNA 159
Synthesis of NBPC (nitrobenzylpyridinium chloride) 159
In Vitro Translation Systems from Higher Organisms 161
Preparing a Wheat Germ Extract 161
Wheat Germ Translation Reaction 163
Preparing a Rabbit Reticulocyte Lysate 163
Micrococcal Nuclease Digestion of Lysate 164
Reticulocyte Translation Reaction 165
Purifying Total and Polysomal RNA 166
Extracting RNA from *Drosophila* Adults 166
Extracting Polysomal RNA from *Drosophila* Embryos 167
Purification of Poly-A^+ (mRNA-Enriched) RNA 168
RNA Size Fractionation by Sucrose Gradient Centrifugation 170
Making the Sucrose Gradients 171
Preparing the RNA 171
Sedimenting and Collecting the RNA 172

Chapter 7 Assorted Laboratory Techniques 173
Glass and Plastic Containers 173
Siliconizing Glassware 174
Washing Pipettes 174
pH Meters 176
Buffers (Tris, Phosphate, Good Buffers, Cacodylate) 177
Beckman Ultracentrifuges 178
Filling and Balancing Tubes 178
Loading Rotors 179
Drawing Figures 179
Slides and Negatives 181
Polaroid Slides 181
Conventional Slides and Negatives 181
Film Sensitometry 183
Autoradiography and Fluorography 184
Dialysis Tubing 186
Distilling Phenol 187
Recovering Used CsCl 188
Sources of Chemicals 189
Hazards and Cautions 190
Desiccators, Preparative Centrifuges, and Vacuum Pumps 190
Chemicals 191
Electrical Hazards and Protections 192

Appendix I Commonly Used Recipes 195

Appendix II Useful Numbers 205

Bibliography 207

Index 213

Chapter 1

Using *E. coli*

Before the rise of recombinant DNA work, much research in molecular biology utilized *Escherichia coli* both as the object of experimental study and as the source of the experimental material. More recently, the diversity of organisms and problems being studied has greatly increased and *E. coli* is being used less as an object of study. However, its use is still appreciable; *E. coli* is often the most appropriate organism for study of fundamental problems or the best source of DNA and proteins necessary for recombinant DNA research. The sections which follow describe the use of *E. coli* both as an object of experimental study and as a vehicle for other studies.

Typically, there are three reasons for growing cells: (1) for the purification of some biomolecule; (2) for genetic manipulations; or (3) for measurements made on growing cells. In these instances, the variables of most concern are the strain, the growth medium including dissolved gases, and the growth temperature.

STRAINS

Strain Purity

Strains are often not what they are thought or claimed to be and instead have unknown growth requirements or other unexpected properties. Also, strains are occasionally contaminated with revertants or with totally different organisms. Therefore any strain to be used should be tested for some of its known properties and, if possible, should also be tested for the properties required by the particular experiments. This testing should usually be done on single colonies that have been obtained by streaking the strain on a YT*

---

* See Appendix I, Commonly Used Recipes.

plate. In this way, several of the isolated colonies can be tested for several of the strain's characteristics. Testing for nutritional requirements is usually sufficient, but if the particular desired mutation reverts rapidly it must also be tested.

Sources of Strains

It has been the practice in molecular biology to freely exchange strains and mutants. However, in recent years many investigators have gone on to different lines of work and no longer maintain strains that they have described in the literature. Fortunately, Dr. Barbara Bachmann of the *E. coli* Genetic Stock Center (Dept. of Human Genetics, 310 Cedar Street, New Haven, Connecticut 06510) has assumed the position of collector and distributor of *E. coli* mutants. She has a large collection of many of the known *E. coli* mutants and has been providing inocula to investigators.

Commonly Used Strains

The K-12 strains commonly used in genetic studies are derived from a culture isolated at Stanford Medical School. The B line is said to derive from Berkeley and B/r was selected from the B line by its resistance to ultraviolet (UV) radiation. Strain B is very sensitive to UV. Strain C grows phage ϕX-174 and lacks a restriction-modification system.

Pedigrees and Genetic Maps

Frequently the mutations necessary for an experiment have already been isolated. Generally it is easier to obtain a strain by requesting it from another investigator and testing it for the properties needed than to isolate, characterize, and test a new mutant. In many cases the exact strain required is not available but a strain with most of the properties does exist and a few simple genetic manipulations will complete the construction of the required strain. Much useful information on existing strains is concentrated in just a few papers. The most useful are "Linkage map of *Escherichia coli* K-12, edition 6," Bachmann and Low, 1980; "Pedigrees of some mutant strains of *Escherichia coli* K-12," Bachmann, 1972; and "*Escherichia coli* K-12 F-prime factors, old and new," Low, 1972. The genetic nomenclature used in these papers and followed as closely as possible in the current literature is described in "A proposal for a uniform nomenclature in bacterial genetics," Demerec et al., 1966.

Strains for Physiological Measurements

Since most strains have been maintained in the laboratory for decades, it is uncertain how much of a strain's current properties are the same as properties measured earlier or in different laboratories. As the complexity of the problem being studied increases, the problems of genetic stability become more severe. Thus, one need not be greatly concerned about evolutionary drift when determining a fairly simple characteristic such as the sequence of a protein. However, if a complicated cellular control system is under investigation, for example, the rate of ribosome synthesis, the regulation of DNA synthesis, or the composition of nucleotide pools, then the origin and possible divergence of strains being used is of great importance.

There are a few rational steps one can take to deal with the problems of genetic instability. In principle, one step would be to use strains which are genetically as close as possible to the original B, B/r, and K-12 strains. If possible, direct descendants of the original strains should be used. If special mutants are required, these are best constructed by moving the desired marker from its strain into a well-characterized standard strain. P1 transduction is the safest way to introduce the marker because little else will be introduced at the same time. If mutagenesis must be used to isolate a particular mutant, light mutagenesis, for example UV treatment, is safest. Nitrosoguanidine mutagenesis, which generally introduces multiple mutations, can create many problems.

Storage of Strains

The shelf life of strains varies widely. Some strains remain viable for months when stored in liquid medium in the refrigerator; others die within only a few days under these conditions. For genetic purposes many strains can be stored for years at room temperature in the dark in small, airtight vials half filled with stab agar. Storing a strain for a series of physiological measurements should be done more carefully. Probably the safest way to insure physiological constancy in a strain over a period of several years is to freeze samples. A single colony can be grown, diluted into medium containing 10 to 50% glycerol, distributed to a number of aliquot vials, and stored at −20 to −70 °C. Such a vial provides viable inocula for about 30 cycles of freezing and thawing.

CELL GROWTH

Properties of Bacteria

Much is known about the growth of *E. coli* and a sizeable fraction of this information is available in *Control of Macromolecular Synthesis,* Maaloe and Kjeldgaard, 1966. This book is notable for its careful analysis of the many factors affecting bacterial cell growth. It is likely that very similar factors will affect growth of other types of cells. Any person handling biological materials should also be familiar with the material presented in *The Microbial World,* Stanier and Adelberg, 1970. Reading this book should prevent such blunders as autoclaving a minimal salts medium in the presence of sugars (this creates toxic sugar phosphates and prevents cells from growing).

A number of specific facts, many of which are not explicitly mentioned in general references on *E. coli* growth, are listed below. Knowledge of these can prevent many of the most frequent mistakes.

Following inoculation a culture generally requires from 1 to 20 h to begin noticeable growth. It then grows exponentially to a density of 1×10^8 to 1×10^{10} cells/ml (Figure 1.1). The growth rate then decreases due to insufficient oxygen, exhaustion of nutrients, or accumulation of waste products. In some cases and particularly in minimal media, cells inoculated at less than a certain critical density will not grow.

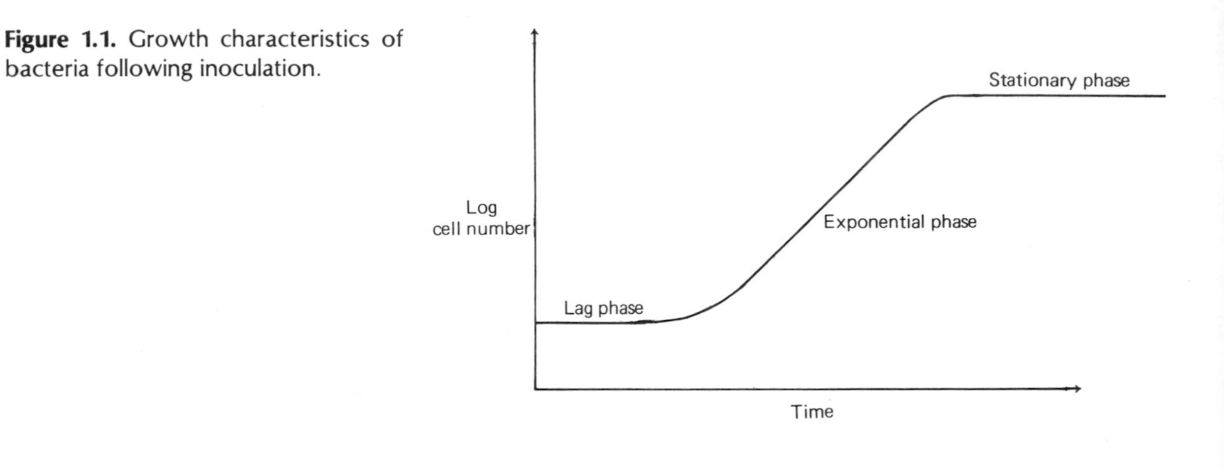

Figure 1.1. Growth characteristics of bacteria following inoculation.

Some compounds and elements, such as SO_4^{2-}, PO_4^{2-}, Mg^{2+}, are well-known trace requirements for growth, but other unknown substances are also needed in trace quantities. Usually a minimal medium contains these unknown substances because they are impurities in the explicitly added chemicals.

Strains possessing "relaxed" control of RNA synthesis (derivatives of strain 58-161) (Stent and Brenner, 1961; Edlin and Broda, 1968; Fiil, 1969), that is most Hfr strains (Bachmann, 1972), accumulate rRNA upon amino acid starvation. Many of these strains do not grow for 12–24 h after they have been diluted from stationary phase. This lag is particularly common with RNase-less strains.

Cells resistant to the phage T1 often do not grow well in minimal medium. This is not true of the old standby T1-resistant strain, C600. Apparently the poor growth of these strains is due to chromic ion interfering with Fe^{2+} uptake. This problem can be overcome by including 0.2% citrate in the medium or by increasing the Fe^{2+} concentration, or both

(Cox et al., 1970; Wang and Newton, 1969; Frost and Rosenberg, 1975). Addition of valine to cells (K-12, not B) growing in a minimal medium that lacks amino acids will cause starvation for branched chain amino acids (Guardiola et al., 1974). This makes it easy to induce amino acid starvation even in a wild-type strain.

Proline at less than 10 μg/ml is degraded about 45 min after its addition to B/r (Schleif, 1967) and, presumably, K-12. Despite this fact, proline is very useful for labeling proteins because (1) the cellular pool size of free proline is small; (2) the proline active transport system is constitutive and very efficient; and (3) proline is at the end of a pathway. Thus proline is especially useful for pulse labeling because it is rapidly incorporated.

Thymine-requiring cells exist as low level, 2 μg/ml, and high level, 15 to 25 μg/ml, requirers (Alikhanian et al., 1966). The most common Thy$^-$ cells are high level requirers and hence a typical Thy$^-$ cell must be supplemented at the higher levels of thymine. As a consequence making highly labeled DNA in vivo with such a strain can be expensive.

Vitamins should be provided at 1 μg/ml, amino acids at 10 μg/ml, and the carbon source at 0.2%.

Some strains clump rapidly. This is useful when large volumes are to be harvested since they will settle, often within an hour, to 5-10% of the original volume.

The streaming pattern seen when cultures are swirled is streaming birefringence, which is due to the rod-shaped cells being aligned along lines of shear in the flowing liquid. The patterns are not seen with spherical cells. The pattern can be used as a rough gauge of the shape and hence the health of the cells. The better cells are growing, the more elongated they are.

Phage Contamination

Phage contamination of bacterial cultures can be a problem and may be particularly troublesome in laboratories where phage are grown in large quantities. Phage contamination usually can be avoided by selecting cells which are phage resistant. Phage T1 presents some special problems because, as mentioned previously, T1^r cells are often sickly (Wang and Newton, 1969). The problem is compounded because T1 is not killed by drying and therefore persists in a laboratory and spreads with dust. The T2 and T4 phages are less of a problem because they are killed by drying. RNA viruses like R17 are an insidious form of contamination because they frequently contaminate small cultures without causing complete lysis and without preventing the cells from reaching stationary phase. The contamination then reveals itself as extensive lysis during growth of volumes of more than 1 liter. Cells free of R17 can be purified from contaminated strains by streaking on YT plates and growing at 33 rather than 37 °C. This appears to be possible because the F pilus to which R17 must adsorb is not appreciably synthesized at the lower temperature. Antiserum to R17 can be made and used at the greatest dilution which still blocks lysis of cultures by phage. Cultures grown in this way generally will not be absolutely free of the phage. Often, contamination problems can be reduced to workable levels by shining several 15-W germicidal lamps in a room overnight and by shining UV on the tops of any carboys before they are inoculated.

Several other practices are commonly used to reduce problems of phage contamination. Any culture which lyses or stops growing for an unknown reason should be autoclaved before opening. This practice will reduce the spread of phage. If a laboratory is contaminated by phage, it is still possible to work with cells if great care is taken. For example, a sterile pipette should be removed from its can without touching it to the other pipettes. This is most easily done by bringing it out along the inside top of the can which presumably has not been touched by fingertips covered with phage. Many problems can be reduced by working with female $T1^r$ strains and, when growing large quantities of phage, by using an amber mutant phage so that escaped phage will be unable to attack Su^- cells that are used for most other work.

Use of Cells for Physiological Measurements

After a purified colony has been tested to demonstrate that it possesses the right nutritional markers, grow it to stationary phase in about 5 ml of YT medium. This culture can be refrigerated and used thereafter for inoculating additional cultures. One should never make a practice of growing cells serially, that is, growing one flask and using this to inoculate cells for the following day's experiment, and so on. This can only lead to disaster, either through contamination or through selection of mutants able to grow more rapidly in the medium. All serious and careful measurements of cells should probably be made on cells growing in a special defined medium (Neidhardt et al., 1974).

As a general rule, any culture to be used in a physiological experiment, that is, for a measurement of the level of any controlled cellular components, should be grown exponentially for at least 5 doublings before the measurements are made. In addition, the cells are best used for measurement at a density of about 1×10^8 cells/ml. At densities higher than this, conditions become nonreproducible primarily due to oxygen limitation, or if reproducible, the reproducibility may be confined to one laboratory. An experiment that requires measurements over a long time period will usually require several dilutions of the exponential phase culture. Dilutions should be into prewarmed and preshaken medium. Often it is convenient to determine a culture's growth rate one day and to dilute it so that cells will be at a density of 1×10^8/ml at a convenient time the following day. One should control all variables such as pH of media, temperature, and agitation of the flasks. For adequate aeration do not use more than 100 ml of medium in a 500-ml flask. Note that rolled tubes do not receive adequate air for most physiological experiments. One additional precaution is to beware of any culture whose growth rate has changed by more than 10% from its norm.

The importance of making critical physiological measurements only on cultures which are known to be in balanced exponential growth cannot be overemphasized. For this reason it helps to make frequent readings of the cultures' optical density (OD) at 450 or 550 nm. Bear in mind, however, that an exponential increase of OD is insufficient to demonstrate balanced exponential growth. A culture diluted from stationary phase can increase OD exponentially for 2 or 3 apparent doublings without any cell division. The cells increase their volume from the small size characteristic of stationary phase to the larger size of normal growth.

Changes in media during an experiment are best accomplished by filtration with a standard Millipore apparatus. Millipore filters of pore size 0.45 μm and 2 cm^2 in area can filter about 3×10^8 cells. The filters should be boiled twice in distilled water before use. This procedure will remove glycerol, detergents, and plasticizers. The filters should be stored in water until used. After collecting the cells on the filter, they can be rinsed with new medium by filtration and then resuspended by shaking the filter in the new medium. A medium change by filtration can be done on 100 ml of culture in 20 s. Centrifugation is occasionally used for media changes, but it takes much longer and the temperature and oxygen shocks could alter the physiology of the cells.

Use of Cells for Genetic Purposes

In contrast to growing cells for physiological purposes, growth for genetic use requires small quantities of cells that are absolutely free of contamination. Initial cultures should be started from purified colonies that have been checked for several of the strain's growth requirements. Tubes containing several milliliters of medium can be grown overnight and stored at 0 to 10 °C for up to a month. Sterile technique must be rigorously followed to prevent any contamination.

All procedures involving selection of a minor subpopulation of cells should include controls that check the purity and reversion rate of strains and the sterility of any phage, DNA stocks, or buffers used.

Cell mating and some other genetic manipulations are usually more efficient when cells are growing exponentially at densities below 2×10^8. Exponential growth for many generations does not seem to be a necessary prerequisite for efficient mating.

Keeping an episome in a strain or curing a strain of an episome is important in many experiments. Plasmids and other episomes are usually lost by segregation. The rate of this loss varies widely and depends both on the host and the episome. Usually, the best way to maintain a culture of cells containing an episome or plasmid is to grow the cells in a medium in which they cannot reproduce unless they have the episome.

The use of recombination-deficient $RecA^-$ cells is occasionally necessary in order to reduce recombination between an episome and the chromosome. $RecA^-$ can be introduced by mating with a $RecA^-$ nalidixic acid-resistant Hfr (see below). The presence of the $RecA^-$ allele can be checked by UV sensitivity.

Markers are frequently moved from one strain to another by mating with either Hfr or F′, or by P1 transduction (p. 29). The male and female strains are grown in rich medium to 5×10^7 to 1×10^8 cells/ml and then 0.1 ml of each is mixed in a test tube and left at 37 °C without shaking. In an interrupted mating, at the appropriate time cells are shaken, diluted, and plated. In the usual F′ transfer, mating can be allowed to proceed for 4 to 5 h before shaking and then either plating or streaking on a plate that selects for the desired strain, the new merodiploid, is performed. Up to 80 to 90% of the F^- cells receive the episome in these extended matings.

It is unnecessary to convert an F^- to an Hfr to map a marker. If an F′ is introduced, generalized chromosomal mobilization will transfer nearly any marker to a new F^- at an efficiency of 10^{-4} to 10^{-6} that of Hfr transfer. This is sufficient for many mapping purposes and it is probably worthwhile to isolate new mutants in F^- strains. It is better to avoid attempts to isolate new mutations in Hfr strains. Although a male strain would appear to simplify subsequent genetics, the mutagenesis necessary to isolate the mutation can inactivate the transfer genes. The result is a strain inactive as a genetic donor or recipient. Mutations in F^- cells into which episomes have been introduced can be located approximately by their linkage to nutritional markers (Low, 1973) and then localized precisely by P1 phage transduction.

In the isolation of new mutants, a heavy nitrosoguanidine treatment yielding 20% survivors will produce 10–50 mutations per cell. Mutagenesis procedures involving heavy mutagenesis and subsequent scoring of clones for the desired property have yielded about 1 out of 2000 cells lacking the enzymatic activity being scored. Thus, if the desired mutant can be easily scored in hundreds of candidates, it is practical to use brute force mutagenesis and scoring to isolate a mutant. $RecA^-$, RNase I^-, DNA polymerase I, and arabinose-binding protein mutants have been found in this way.

Growing Cells for the Purification of Molecules

The strain and growth conditions should be chosen for optimum yield of the desired molecule. Frequently it is desirable to use DNase- or RNase-negative strains in addition to special mutants overproducing the desired molecule.

A colony from a YT plate can be grown overnight in a test tube. This can be stored in the refrigerator and 0.1 ml used to inoculate 4-liter flasks each containing 1 liter of medium. The largest yields of protein, although not necessarily of a given enzyme, are obtained by growing cells at room temperature rather than at 37 °C. Dow antifoam A, at 2 ml/10 liters, reduces foaming. Cells are usually chilled before harvesting, although this is often unnecessary. Centrifuged cells can be used immediately or stored frozen at −70 °C.

MEASURING CELL DENSITY

Most experiments with growing cultures require measurement of cell density to an accuracy of at least 10%. This can be accomplished with a Petroff-Hauser cell chamber (Arthur

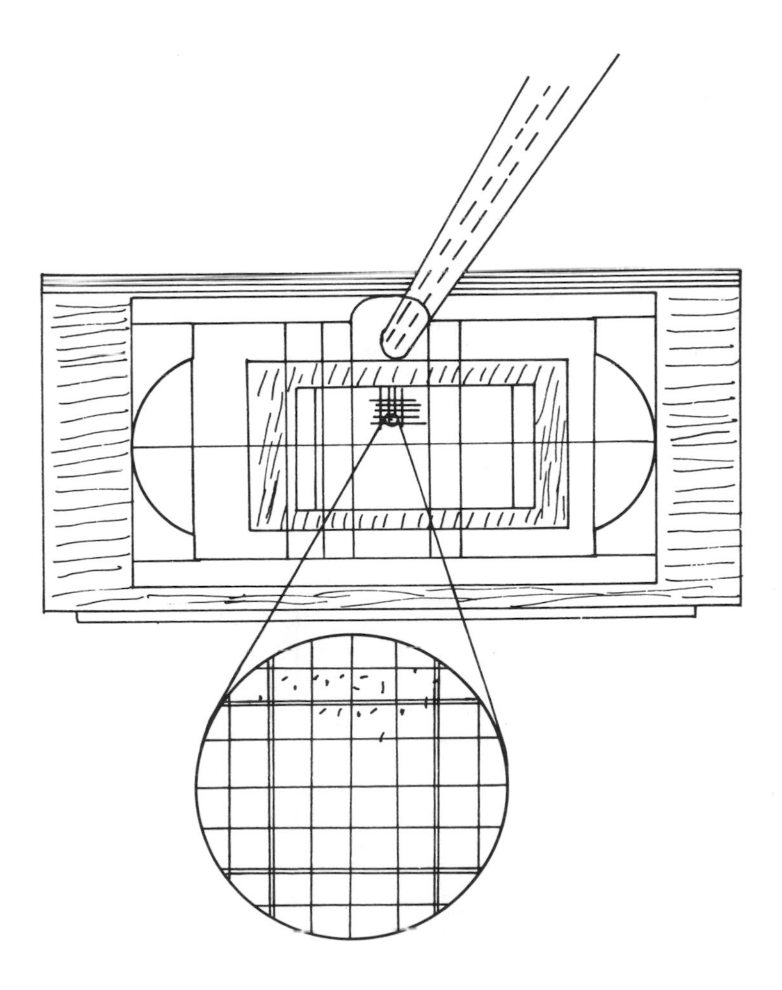

Figure 1.2. A Petroff-Hauser cell counting chamber and a representative view of the chamber counting area. ~ 500 ×.

H. Thomas Co.) and a phase contrast microscope using a magnification of about 500. Place approximately 0.1 ml of cell suspension at the edge of the coverslip; and permit capillary action to draw the liquid into the space between the coverslip and the slide (Figure 1.2). If you add too much medium it will seep into the neighboring groove and the cells will flow rapidly across the microscope field, making them impossible to count. When this happens it is best to dry the cell chamber and start over. If you add just a little too much medium and it forms a small drop beside the coverslip, the cells will also flow rapidly. Use a tissue to draw off the excess medium and the cell flow will then stop. The counting chamber is constructed so that 5×10^{-8} cm^3 of liquid will be contained between the slide and coverslip within an area bounded by one of the small etched squares. Thus, one cell per square corresponds to 2×10^7 cells/ml.

Cell numbers are also measured with a spectrophotometer since, up to an apparent optical density of about 1.5, the absorbance is proportional to cell density. The correspondence between absorbance and cell number can be measured for any particular strain in a particular medium. Either a Petroff-Hauser counting chamber or calibrating by diluting and plating viable cells can be used to establish the conversion factor.

An optical density measurement of a cell suspension with a spectrophotometer is not a measure of absorbance, but rather a measure of scattering. Light is scattered out of the collimated beam by the cells and does not reach the photomultiplier (Figure 1.3), whereas with a colored solution, the light is absorbed by the solution and therefore does not reach the photomultiplier. Since the light is scattered by cells, the amount reaching the photomultiplier is a function of the cell density and of the distance between the cuvette and the photomultiplier. For an absorbing solution, the light reaching the photomultiplier is independent of this distance.

The range of response linearity ought to be measured, because once the bacterial concentration is sufficient to scatter a large fraction (approximately 90%) of the light out of

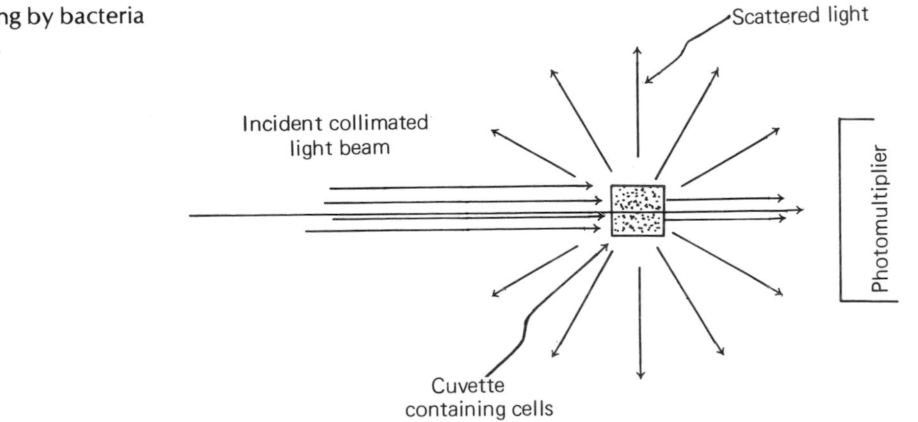

Figure 1.3. Light-scattering by bacteria in a spectrophotometer.

the main beam, further increases in bacterial density will lead to little increase in the indicated optical density. On Zeiss instruments nonlinearity is less than 10% when the indicated OD is below 1.5. For this reason, dilutions are required for accurate readings when the culture OD is more than 1.5. Typically, on a Zeiss spectrophotometer, an absorbance of 1 at 550 nm occurs at 2×10^8 cells per ml.

GROWING LARGE QUANTITIES OF CELLS

Inevitably large quantities of cells must be grown to permit purification of proteins or nucleic acids. Fortunately, expensive equipment is unnecessary and cells adequate for most purposes can be grown in 5-gallon Pyrex carboys fitted with a means of providing aeration and any necessary heating or cooling.

Up to 30 liters of cells may be conveniently grown at one time in two 5-gallon Pyrex carboys. Carboys of this size are about the largest that can be safely handled. The carboys, containing medium plus aeration and heating plumbing, can be autoclaved in standard large-sized autoclaves. After cooling, the carboys can be used on a bench top to grow cells. Sterile aeration is provided by passing "house" compressed air through sterile cotton. The carboys can be operated at most desired temperatures utilizing a circulatory waterbath to pump heating water through a bent exchanging coil in the carboys. If frequent use is to be made of the carboys, it is convenient to have metal collars for them. In addition to an inoculation port, the collar can contain the bent exchange coil and the aeration tubing. For occasional use carboys can be grown in a warm room and require nothing more specialized than gas dispersion stones connected to a sterile air supply with a rubber hose. They can be held in place with cotton packed into the neck of the carboy.

PROCEDURE

To prepare carboys, add a couple of liters of water to each carboy, then add the desired nutrients and thoroughly stir. Add the remaining water and about 3 ml of Dow Corning antifoam Y-30 emulsion. This is most easily dispensed from a plastic squeeze bottle. Put on a metal cap containing the air hoses, cotton sterilizing filter, and gas dispersion stones (Fisher No. 11-139B) (Figure 1.4). Tighten the screws on the collars, but do not make them snug. Leave considerable slack because the glass may expand more than the metal collar during autoclaving. Place a 100-ml glass beaker over the sample port. Autoclave the carboys with the filter and inoculating tubes connected, 60 min sterilizing and 60 min exhaust. The total time in the autoclave with cooling will be approximately 3 h. Be extremely careful while handling hot

Figure 1.4. A collar assembly for a carboy. The figure shows the inoculation port which is normally covered with a sterile beaker, the air in port, and the circulated heating water ports. The heating water ports are connected to a coil of stainless steel tubing narrow enough to pass through the neck of the carboy. The air in port is connected via rubber tubing to gas dispersion stones that are located near the bottom of the carboy to provide complete aeration.

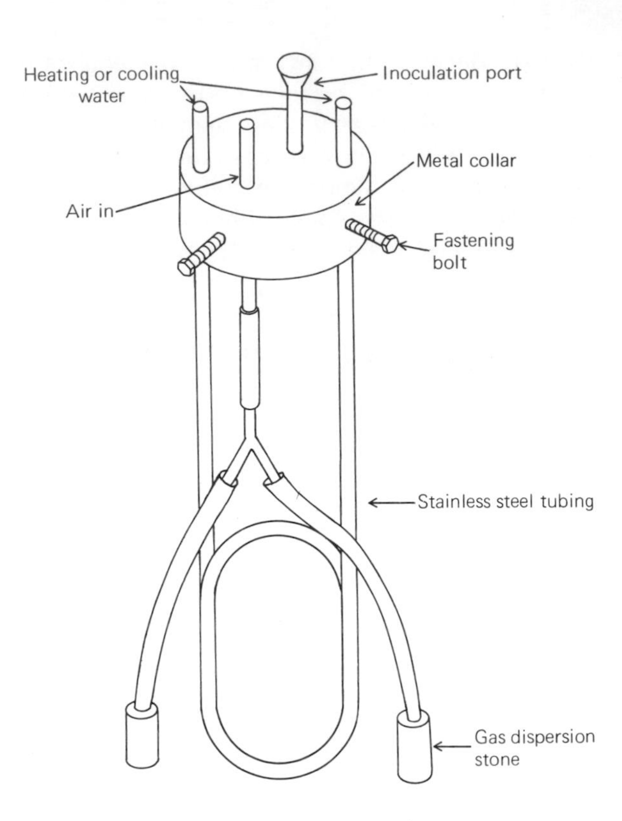

carboys. Being burned with 15 liters of boiling water can be lethal. Bring the carboys back to room temperature by circulating tap water through the coils at a rate of about 100 ml/min. They will cool faster if they are stirred by a slight flow of air through the dispersion stones. Bear in mind, however, that the carboys are not absolutely sterile at this point. They can sit at room temperature overnight without aeration. However, if carboys are maintained at 37 °C with aeration but no inoculation for 12 h, about one out of three carboys will be found to be substantially contaminated.

The carboys can be inoculated through the sample ports or by removing the cotton plug used to seal the top if a metal collar is not used. Often it is convenient to have the carboys inoculated in the middle of the night. This can be accomplished automatically by using a 24-h timer. This can turn on the circulating water bath which heats the carboys, and at the same time open a solenoid valve in the air line. As shown in Figure 1.5, if the inoculum is contained in a U-tube between the air filters and the carboys, when the air is turned on the inoculum will be blown into the carboys, thus inoculating them. Thereafter the sterile air will be blown through the U-tubes and gas dispersion stones in the carboys.

A Sears 24-h timer works adequately, and a Haake heater type FE has sufficient capacity to keep two carboys at 37 °C in a room at 20 °C. A satisfactory solenoid valve is ITT type S301AA02N3BE1. The specifications state that this valve will control 100 lb/in.^{2} through $\frac{9}{16}$-inch ports with a 120-V, 60-cycle supply. Evaporation from the carboys promoted by the aeration slightly cools them, and it is usually necessary to set the temperature of the circulating water bath about 1.5 degrees higher than the temperature desired in the medium. The cotton filter and inoculation U-tube can both be made from the outer portion of a 3 × 50 cm thistle tube condenser.

Cells from a carboy may be harvested in 15–30 min with a DeLaval Gyrotester centrifuge (Poughkeepsie, New York). Not only is this a simple device to use, but its price is reasonable. Harvesting times longer than 30 min are required when flow rates are reduced to allow complete chilling of the cells before they enter the centrifuge. The plumbing is arranged so that a long glass tube reaches to the bottom of the carboy. Clear plastic tubing may be used to connect this tube to a coil of stainless steel tubing (5 m × 1 cm) immersed in ice water if cooling during harvesting is necessary. Finally, connect the other end of the coil to the top of the gyrotester as shown in Figure 1.6. Place the coil of steel tubing in a plastic tray filled with ice water, and put a circulating pump into the tray, and arrange its inlet and outlet so that the pump rapidly circulates ice water through the tray. The centrifuge can rest on the floor with its outlet leading to a bucket. To start the siphoning of medium from the carboy, which is placed on a lab bench, connect the hose that enters the gyrotester to a water tap. Run water backwards through the system, flushing out the air until water just enters the carboy. Disconnect the hose from the tap and quickly put it back into the gyrotester. With a pinch clamp anywhere in the system, adjust the flow rate to between 500 and 1000 ml/min.

Figure 1.5. Schematic of carboy aeration, heating, and inoculation. The figure shows the 24-h timer into which the solenoid air valve and the circulating water bath are plugged. Also shown is the pathway of air flow through the solenoid valve sterilizing cotton filters, inoculation tube, and into the carboys. The heating water is circulated by the circulating water bath through the coil contained within the carboy.

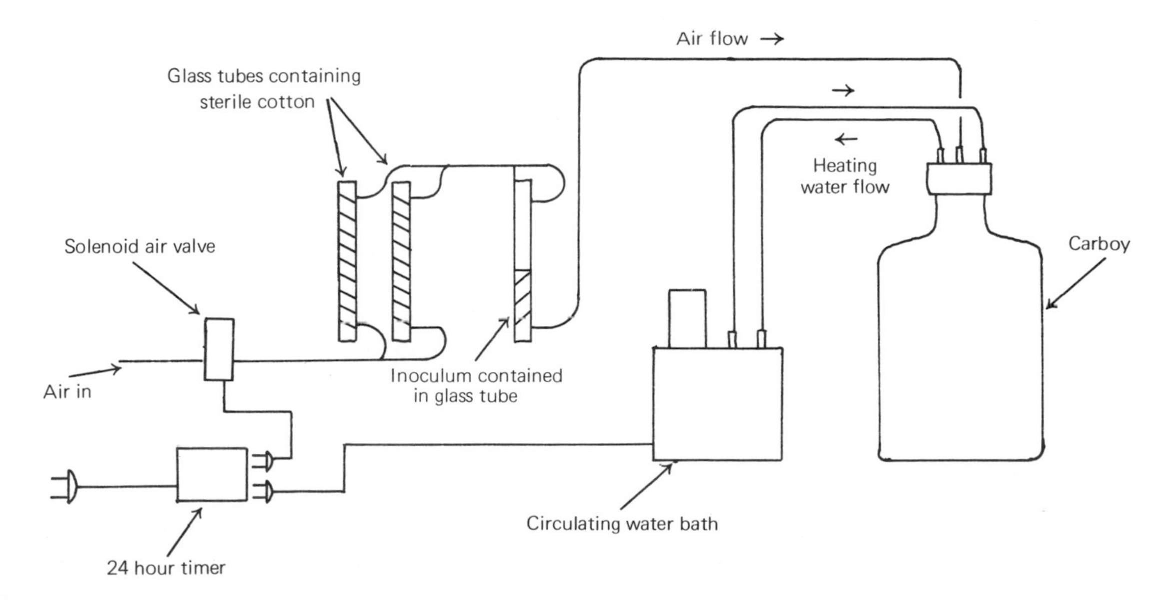

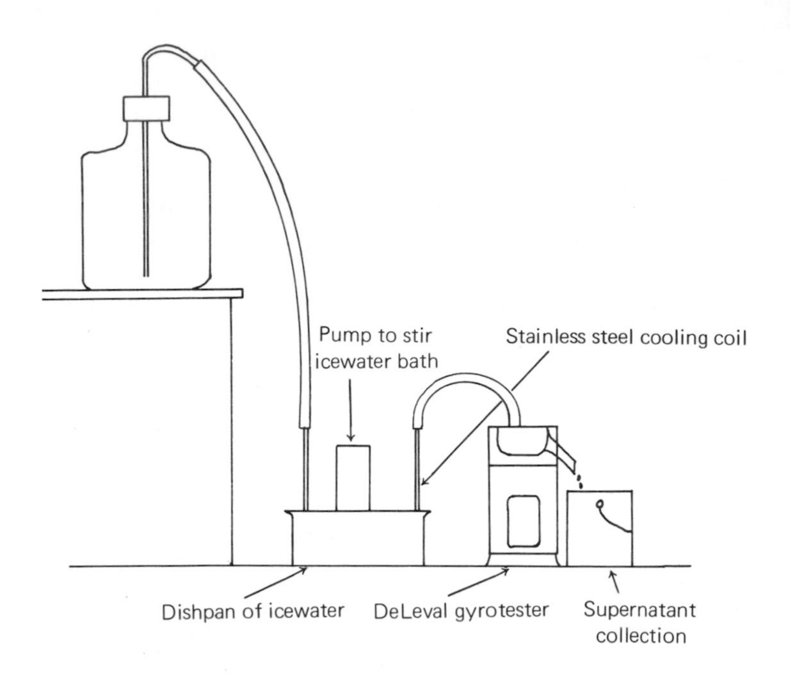

Figure 1.6. A convenient arrangement for harvesting cells from a carboy. Gravity flow of cells proceeds through the stainless steel cooling coil and into the DeLaval gyrotester (centrifuge). The waste supernatant is collected in a plastic pail.

The capacity of the gyrotester rotor is about 100 g of cells. Usually 75–100 g of cells are harvested from each carboy (1 OD [Zeiss] at 550 nm corresponds to approximately 1 g of cells/liter).

When assembling the rotor of the gyrotester, note that each of the internal cones is numbered. Place them in order in the rotor. Tighten the top of the rotor with a strapwrench, making sure that the index marks align. The top of the shaft of the gyrotester should be kept lightly greased. Place the rotor on the shaft and then remove and replace it several times, using the thumb of your other hand as an index of how far the rotor sits down on the shaft. With 2 min of experience, you will be able to tell when the rotor has been properly seated on the shaft. The only critical operation in the entire centrifugation procedure is seating the rotor.

If the motor is run when the rotor is improperly seated, extensive damage can occur to the rotor, shaft, and the bearings. If the rotor beings to bounce when the motor is turned on, then it is not properly seated. If the motor is promptly turned off, little or no damage will occur and the rotor probably will have been seated in the process.

Occasionally the stainless steel bowls that surround the rotor and collect the liquid as it comes out will scrape the rotor as it is running and produce a terrible howl. This can be corrected without stopping the rotor by pushing and pulling on the bowls until the noise stops.

After centrifugation, the rotor is easily cleaned with hot water. Note that cells often stick under the cone which is attached to the rotor top. They can be removed by first removing the cone, which is held in place by a left-handed nut on the top of the rotor. When reassembling, take care to match the index marks.

OPENING CELLS

We describe several methods because different quantities of cells are best opened by different methods. With any of these methods you should use a buffer that is kind to the molecule you wish to purify. Many enzymes can be purified from cells opened in 0.01 M Tris-HCl, pH 7.5; 0.1 M KCl; 0.01 M $MgSO_4$; 10^{-4} M dithiothreitol; and 10% glycerol.

Sonicating

Up to 1 g of cells (approximately 10^{12} cells) can be easily opened by sonication. Care must be taken not to let the temperature rise too high during the procedure. Cells can be sonicated at any concentration up to approximately 2×10^{11}/ml. Sonication is often used when many samples are being assayed for a particular enzyme. If the sonicator possesses a tuning control, it is best to attempt to obtain a "frying egg" sound from the sonicated solution. Usually it is most effective to place the bottom of the probe just beneath the surface of the cell suspension. Relatively few enzymes are sensitive to sonication.

Grinding in Alumina

Grinding in alumina is convenient for 0.5-100 g of cells. One weight of cells, paste or frozen, and 2.5 weights of levigated alumina are ground in a precooled mortar to a smooth consistency like cake icing, having first passed through a stage where the creamy mix snaps like bubble gum. After grinding, add buffer slowly with mixing. Usually 2 weights of buffer are added. Add DNase I (1-10 μg/ml) and incubate the mixture at 0-4 °C for 15 min before spinning out the alumina and debris at $5000 \times g$ for 10 min. This is a good way to open cells for the extraction of labile proteins.

Grinding with Glass Beads

Grinding with glass beads is convenient for larger scale work, involving 100-2000 g of cells.

WASHING GLASS BEADS

1. Use Superbrite 100 beads, produced by Minnesota Mining and Manufacturing Co. New beads should be soaked overnight in concentrated HCl.
2. Rinse the beads with water in a large sintered glass or Buchner funnel.
3. Resuspend and stir with 2 vols of 0.5 N HCl.

4. Rinse with distilled water until the pH is 4.
5. Wash with 2 vols of 0.5 N NaOH.
6. Rinse to pH 8 with double-distilled water.
7. Suck the beads as dry as possible on the funnel.
8. Dry overnight in a shallow baking pan in a drying oven.

GRINDING

1. Break the frozen cells into cherry-sized pieces and add 2.5 weights of prechilled washed glass beads and 1 weight of breaking buffer.
2. Grind in a water-jacketed Waring blender. Cool the jacket with circulating ice water. Grind 5 min at low speed and then 10 min at high speed.
3. Add 1 mg DNase I for each 100 g of cells and grind 1 min at low speed.
4. Let the beads settle for 20 min at 4 °C and decant the supernatant.

Opening Cells with Lysozyme

Cells can also be opened with a combination of lysozyme to digest cell walls and sodium deoxycholate to disrupt cell membranes. This procedure is useful when the molecule you are extracting from cells can survive an absence of magnesium ions. It is necessary to remove free magnesium ions in order for lysozyme to function. The following procedure is a modification of one described by Burgess and Jendrisak (1975):

Buffer is 0.05 M Tris–HCl, pH 7-9, 5% (v/v) glycerol, 2 mM EDTA, 0.1 mM dithiothreitol, 0.233 M NaCl, and 130 μg/ml of egg white lysozyme (made fresh the day of use). During the early stages of purification many enzymes are susceptible to proteolytic digestion and the inclusion of $PhCH_2SO_2F$ (phenyl methyl sulfonyl fluoride [PMSF]), an inhibitor of the "serine proteases," to 23 μg/ml, often dramatically increases recoveries. Break 1 weight of cells which have been frozen at −70 °C into pieces smaller than cherries and grind in a Waring blender with 3 weights of grinding buffer for about 3 min, so that they warm to 0–4 °C. After about 20 min at 0–4 °C, add a sodium deoxycholate solution with stirring to give a final concentration of 0.05%. Blend the mixture at low speed to mix and leave at 0–4 °C for 20 min before mixing at high speed to shear DNA.

RADIOLABELLING CELLS

When labelling cells with a radioactive precursor to determine the relative rate of synthesis of some macromolecule, it is necessary that labelling conditions do not disturb cell growth. Particularly when labelling with $^{32}PO_4^{2-}$ or $^{35}SO_4^{2-}$, care must be taken to insure that cells have adapted to growth in medium containing the lower concentrations of SO_4^{2-} or PO_4^{2-} that are necessary for efficient labelling. Occasionally it is necessary to reduce the ion concentrations in 3-fold steps, each step adapting cells by a 10-fold growth.

Several amino acids can be used for efficient labelling of protein. As explained above, proline has several very desirable attributes: It is taken up from very low concentrations, i.e., 10^{-7}-10^{-8} M, in the medium; its transport is constitutive; the pool size of free proline inside cells is about a 10-sec supply; and proline is not normally converted to other amino acids. Proline can be used to label in a minimal medium, in a minimal medium supplemented with a synthetic amino acid mix, or with amino acids plus nucleic acid precursors.

Proline can be used for very short pulse labelling of cells, for example 20 sec, since uptake begins within 5 sec of its addition and further incorporation can be stopped in 10 sec by adding excess nonradioactive proline (Schleif, 1967). Other amino acids may have these attributes as well. Uracil is frequently used to label RNA, but it also is incorporated into DNA, so be wary.

NITROSOGUANIDINE MUTAGENESIS

Nitrosoguanidine (NTG) mutagenesis is the most convenient method for producing extensive mutagenesis when rare mutants are being sought (Adelberg et al., 1965). It should be remembered that, along with the desired mutation, your new mutant strain is likely to contain several other mutations. Also, bear in mind that nitrosoguanidine is a potent skin irritant and a carcinogen.

Nitrosoguanidine primarily mutagenizes the growing points of DNA (Cerdá-Olmedo et al., 1968). Thus, the mutations introduced into any single cell will be concentrated in the regions of the growing points of that cell at the time of mutagenesis (Oeschger and Berlyn, 1974).

PROCEDURE

Centrifuge 5 ml of an exponentially growing culture, wash it in TM (0.508 g maleic acid, 0.605 g Tris base, 100 ml dH_2O adjusted to pH 6.0 with NaOH), and resuspend in 10 ml of TM. Add 0.6 ml of 2 mg/ml NTG that has been freshly dissolved in TM. Allow the culture to stand at 37 °C for 30 min without shaking. Wash the cells once and suspend them in the desired medium. If the desired mutation does not appear, vary the time that the cells are incubated with NTG and monitor the stimulation in mutation frequency and the killing. Killing ought to be between 10 and 90%. A simple marker to use when following mutation frequency is valine resistance. This procedure ought to increase the frequency of valine-resistant cells about 300-fold.

PENICILLIN SELECTION

Some mutants may be isolated by selecting cells which do not grow under certain conditions. The penicillin treatment allows the selection of such cells by killing growing cells. The method is adapted from Gorini and Kaufman (1960).

SELECTING FOR Arg^- CELLS

1. After mutagenesis, grow cells through at least 5 doublings or to stationary phase in minimal medium plus arginine.
2. Spin down cells, wash, and resuspend at about 1×10^8cells/ml in minimal medium containing 0.2% glucose as a carbon source, lacking arginine but containing 20% sucrose to osmotically protect the spheroplasts which will be formed, and 0.01 M $MgSO_4$.
3. Incubate the culture until cells are growing well and the OD has increased by a factor of 4. It is *very important* to have cells growing well for the next step, so it is worthwhile to make frequent OD readings and to plot them on semilog paper.
4. When cells are growing, add penicillin to give 2000 units/ml. It may be better to use a large flask so that the liquid is less than 0.5 cm deep and not to shake the flask during the penicillin treatment.
5. Continue the penicillin treatment for about 90 min or about the doubling time of cells in the sucrose-Mg^{2+} medium. At the end of the treatment, 50% or more of the cells should be protoplasts and will look ruffled and rounded, as shown in Figure 1.7, rather than smooth and elongated.

Figure 1.7. Typical appearance of *Escherichia coli* cells before and after penicillin treatment in the presence of sucrose to prevent osmotic shocking of the spheroplasts.

6. Spin out cells and rinse once in the same medium lacking sucrose. There should be much lysis and clumping in the culture. The cells can then be grown in liquid medium or plated.

COMMENTS

1. Serial penicillin treatments are effective. They appear to be most efficient when cells are not grown to stationary phase between treatments. Penicillin-resistant mutants are selected by each treatment and result in a 10-fold decrease in the enrichment efficiency with each successive treatment cycle. Since a typical enrichment is 10^3-10^4 on the first cycle, 3 penicillin treatments are usually the useful limit.
2. When selecting for nonauxotrophic mutants, a rich medium without sucrose can be used. Results seem better when $MgSO_4$ is added, however.
3. Although the enrichment provided by the penicillin treatment is greatest when sucrose is present because the latter prevents lysis and hence crossfeeding, often it is such trouble to induce the cells to grow in this medium that it is faster just to perform the penicillin treatment on cells growing in M10 medium* at the time that their density has reached 10^7cells/ml.
4. Occasionally it is necessary to perform a selection on cells which are penicillin-resistant. This situation can arise if the cells contain a plasmid carrying resistance determinants to the penicillin analog, ampicillin. A cycloserine selection may then be used essentially as described by Miller (1972).

CURING CELLS OF F-FACTORS

During genetic constructions it is often necessary to remove an F-factor from a strain. No single technique has proven universal; however, it is rare indeed to find that none of the following techniques work. Several of these methods appear to depend on preferentially inhibiting the replication of F-factors. As a result some of the growing cells find themselves without an F-factor. The other methods we describe appear to depend on selectively killing cells which contain an F-factor.

Curing with Acridine Orange

Acridine orange curing is the most reliable method. The following version follows that described by Hirota (1960).

Make an acridine orange solution, 500 μg/ml in H_2O, autoclave and store in the dark for up to 1 year. The medium is 10 g peptone and 10 g yeast extract per liter adjusted with NaOH to pH 7.6 (pH is important) or YT adjusted to pH 7.6

* See Appendix I.

with NaOH. Grow the culture to be cured overnight in YT. Inoculate about 10^4 cells into a series of tubes containing 0, 10, 20, . . . , 100 μg/ml of acridine orange. Incubate these cultures overnight at 37 °C in the dark. Take or streak out cells from the tube showing growth with the highest drug concentration.

The curing rate can be as high as 100%, but is more frequently 10%. Some episomes cannot be cured by this method.

Selecting Spontaneously Cured Cells

Since most episomes are lost spontaneously from cells, many cultures contain cured cells at a level of about 10^{-4}. Cultures can be grown to stationary phase several times, and a penicillin selection performed to select for loss of a nutritional marker carried on the episome but not on the chromosome. An example is the loss of an F′ leu episome from an F′ leu/leu$^-$ strain, in which case leu$^-$ cells should be selected.

Curing with Sodium Dodecyl Sulfate

The discovery and theory of this approach are described in Inuzuka et al. (1969). Apparently male cells are selectively killed by the sodium dodecyl sulfate. Inoculate about 100 cells into tubes with increasing concentrations of sodium dodecyl sulfate ranging from 0.5 to 10%. After 2 day's growth plate out cells from the tube with the highest SDS concentration showing detectable growth.

Curing with High Temperature

It is reported that episomes are lost from cells grown at high temperatures (Stadler and Adelberg, 1972).

MAKING CELLS STREPTOMYCIN-RESISTANT

Streptomycin resistance is a convenient genetic marker. Its particular value lies in its ability to kill donor cells in plate mating without interfering with their ability to mate (Miller, 1972).

The target of streptomycin in sensitive cells is ribosomes, and killing appears to result from the accumulation of incorrect translation products (Davies et al., 1964). One should be aware that such translation products will be synthesized in diploid streptomycin-sensitive/streptomycin-resistant cells and thus such cells will be streptomycin-sensitive.

To isolate resistant mutants, grow cells to stationary phase in YT. Spin down 5 ml of cells and resuspend in the liquid which remains with the pellet. Spread all of this mixture on a YT plate on which 0.2 ml of 1% streptomycin has been spread. In 2 days about 20 resistant colonies will have grown up. Purify resistant cells by streaking resistant colonies on a YT plate. Test the purified strain for streptomycin resistance. Typically about 20% of these candidates are streptomycin-resistant.

CROSSING *rec*A INTO CELLS

The very useful recombination-defective *rec*A mutation (Clark and Margulies, 1965) can be conveniently crossed into a strain by mating. Use an Hfr strain (Miller, 1972) which transfers the nalidixic acid-*rec*A region of the chromosome early and select for nalidixic acid resistance. The *rec*A recombinants can be recognized by their smaller colony size. They can then be tested for their *rec*A character by assaying UV sensitivity.

Grow the male and female cells to be used for mating exponentially to a concentration of about 5×10^7 cells/ml and then mix 0.2 ml of each and incubate at 37 °C without shaking for several hours. Vortex the mating mix, add 2 ml of YT medium, and grow the cells overnight. This growth permits recombination and gives time for the recessive Nal^r marker to be expressed. Streak or spread the overnight culture of these cells at various dilutions (0.1 ml of a $1/100$ dilution usually gives 200 colonies) on appropriate selective plates. After overnight growth the plates should possess two sizes of colonies. The smaller colonies are much more likely to be Nal^r and these should be purified by streaking on YT plates.

The selective plates are made by adding 0.1 ml of nalidixic acid (Sigma) (4 mg/ml in 1 M NaOH) to a YT plate and allowing the solution to soak in for 15 min. An appropriate agent to select against the donor must also be provided. Phage T6 or streptomycin are convenient and effective if the female and male have the appropriate resistances before the mating.

Purify the candidates and test for nutritional requirements. Assay for the *rec*A marker by UV sensitivity. Grow a culture of the candidate and a culture of appropriate controls to stationary phase in YT and then spread in streaks across a YT plate. Normally 5 streaks will fit on a plate. Irradiate half of each streak for 10 sec, 40 cm away from a standard commercial 15-W germicidal lamp. Shield the other half of the plate with a piece of cardboard or paper. After 6–8 h of growth at 37 °C all cells in the area not exposed to UV will have grown. The *recA*-containing cells will show little or no growth where exposed to UV, whereas Rec^+ cells are oblivious to this dose.

PHAGE P1 TRANSDUCTION OF GENETIC MARKERS

About 1% of the phage particles produced by growing P1 on a normal host contain host chromosomal DNA segments of about 100 kb. Infection of a second strain with such a lysate will produce the following responses: some cells will produce more P1; some cells will be lysogenized by P1; and some cells into which only chromosomal DNA from the previous host has been injected will recombine this DNA into their chromosome. It is the latter property of moving chromosomal markers between strains, independent of their sex, which makes P1 so useful. In addition to easily moving markers, P1 is valuable in genetic mapping. It can be used to check the approximate map location of genes by showing whether or not two markers can be transduced together (cotransduced). It can also be used in fine structure genetic mapping (Lennox, 1955).

PROBLEMS AND LIMITATIONS

1. $RecA^-$ cells are transduced at 10^{-3}–10^{-4} the efficiency of Rec^+ cells.
2. The yield of P1 on $RecA^-$ hosts is about 10% that obtained on Rec^+ cells.
3. P1 carries a restriction-modification mechanism and therefore cells lysogenic for P1 degrade DNA from any source that is not also lysogenic for P1 and therefore appropriately modified. This restriction prevents mating or transduction studies of a P1 lysogen. Hence some care must be taken to prevent formation of lysogenic transductants.
4. Most methods for purifying P1, including the method described below, yield from 5×10^8 to 5×10^{11} PFU/ml. This variability must be kept in mind when performing experiments.
5. P1 does not form good plaques on *E. coli,* and it is usually titered on a laboratory strain of *Shigella* where it forms slightly better plaques.

GROWING THE PHAGE

1. Grow host cells overnight in YT.
2. Add sterile 1 M $CaCl_2$ and $MgSO_4$ to give a final concentration of 0.01 M $CaCl_2$ and 5×10^{-3} M $MgSO_4$. If a precipitate forms immediately, repeat and use less $CaCl_2$.

3. To 0.1 ml of cells add 0.1 ml of phage containing 10^5–10^6 P1 particles. Incubate 15–20 min at 37 °C.
4. Add 2.5 ml soft agar (48 °C) containing 0.01 M $MgSO_4$ and 0.005 M $CaCl_2$. Pour on YT plates containing 0.01 M $CaCl_2$. Freshly poured 20-mm-thick plates seem to work a little better than 15-mm plates which have been stored at 4 °C for a few months, but the difference is rarely important.
5. Grow 4.5–7 h at 37 °C until there is confluent lysis of cells. Compare the resulting plate to a plate which received cells but no phage. Sometimes the cells do not appear to have lysed but a high titer will result anyway.
6. Add 5 ml of YT containing 0.01 M $MgSO_4$. Break up the top agar and scrape it into a Sorvall tube. Add 10 drops of chloroform, vortex, let sit about 5 min, vortex again, and then spin at 10,000 × *g* for 20 min. Store the supernatant in a refrigerator. Some stocks will retain their titer and transducing ability for years, whereas others are spent in 2 weeks. About 10% of the P1 stocks made in this way have such a low titer that they are unusable.

TITERING

For many applications it is unnecessary to titer and more practical to do the transduction with different dilutions of the phage.

1. Dilute the P1 to the appropriate concentration in YT medium.
2. Add $CaCl_2$ to give a concentration of 0.01 M in exponentially growing *Shigella* that are not above 1 × 10^8 cells/ml *(important)*. Do not chill the cells on ice. Add 1 ml of diluted P1 to 0.1 ml of these cells. The titering strain is simply known as "*Shigella* for titering P1" and is passed from lab to lab.
3. Let sit 15 min at 37 °C for adsorption, add agar, and then pour. Fresh plates make the plaques easier to see.

TRANSDUCING

1. Grow cells overnight to stationary phase in YT medium.
2. Add $CaCl_2$ to 0.01 M and add 0.1 ml of P1 to 0.1 ml of cells. Let the phage absorb for 15 min at 37 °C, and then spread 0.1 ml on selective plates. In 1 to 2 days 100–1000 colonies will appear. Transducing with serial 10-fold dilutions of the P1 lysate is the most efficient approach to obtaining an appropriate multiplicity of infection. Take a transductant from the plate which received the fewest phage yet also contains at least 10 times as many transductants as revertants. Measure the reversion frequency by spreading the same number of cells on a selective plate with no P1. It is wise to check the P1 stock for sterility by spreading 0.1 ml of it on a selective plate.

3. A transductant should be purified on YT and can be checked for being nonlysogenic by streaking a growing culture against a dried streak of P1 at 2 $\times$ 10^7 pfu/ml on a YT + $CaCl_2$ plate.
4. Lysogenization of transductants can be prevented by using a low multiplicity of infection and removing $CaCl_2$ after phage absorption. $CaCl_2$ is removed by having 0.01 M Na^+-citrate (neutralized to pH 7.2) in the selective plates.
5. Note that to transduce a recessive marker such as streptomycin resistance there must be growth in liquid, so that the recessive and dominant markers can segregate before the cells are spread on selective plates. After the 15-min absorption, add 5 ml of YT containing 0.01 M Na^+-citrate, grow cells to 10^9/ml, and then spread 0.1 ml on selective plates.

LARGE-SCALE GENETIC CROSSES

Many genetic crosses may be performed at the same time by spotting the males or females on a plate, growing them up, and replicating onto another plate which has been spread with 10^9-10^{10} cells of the other mating partner. This method is convenient if many males are to be crossed with one female or vice versa. Since the efficiency of mating is low, the method is unsuitable if rare recombinants are being sought. This replica plating method is most useful when transfer of an episome always leads to recombination.

If recombinants are rare or if many different males and many different females are to be crossed, growth and mating of the cells in liquid, followed by spotting onto solid selective medium, is the most efficient procedure (Schleif, 1972). In order to have all cells at the proper density for mating and to simplify dilutions of the cells after overnight growth, the cells are grown overnight in a medium in which nutrients are exhausted when cells reach about 1 $\times$ 10^8 cells/ml. In the morning additional required nutrients are added and the cells are grown for several more hours. Cells are pipetted into wells of sterile polypropylene blocks, permitted to mate for several hours, and then spotted onto selective medium in Pyrex trays. Several different trays may be spotted from the same block of mating mixes.

MATERIALS NEEDED

1. Mating blocks: sterile polypropylene blocks, 20 $\times$ 30 $\times$ 2 cm, with loose-fitting stainless steel covers. The blocks are drilled with 1-cm holes that are 1.25 cm deep, as shown in Figure 1.8. The mating blocks are sterilized by autoclaving and any condensed water which remains in them is shaken out while they are still hot.

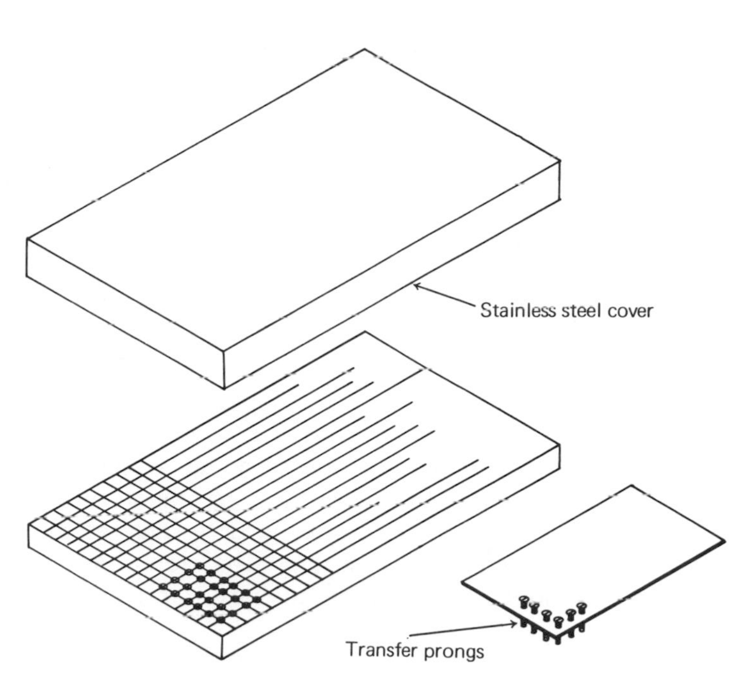

Figure 1.8. A polypropylene mating block, a cover which fits over the block, and the transfer prongs with projecting machine screws.

2. Transfer prongs: an aluminum plate with 0.4-cm machine screws projecting at least 2 cm. The ends of these screws must be flush and flat. They are arranged in an array to fit one quadrant of the mating block.
3. Pyrex trays: 23 × 36 × 4 cm
4. Growth mix: 20× concentrated M10 medium, 5 ml; H_2O, 95 ml; YT medium, 4 ml; $MgSO_4$, 0.01 M, 1 ml; autoclave.
5. Mating mix: Adjust the pH of YT medium to between 6.8 and 7.0 by dropwise addition of 1 N NaOH. Autoclave and afterward add 20% glucose to a final concentration of 0.2%. The pH adjustment and the glucose addition increase the mating efficiency up to 10-fold and may be omitted for many applications.
6. Mating tray medium for 3 trays: Pour the medium into 3 sterile Pyrex trays and permit the agar to solidify before covering with aluminum foil. Use the trays the day they are poured. Airborne contamination cannot be totally avoided in this procedure. However, it rarely causes problems since *E. coli* grows faster than almost all contaminants.

   See "Commonly Used Recipes":

   A component of minimal plates, 500 ml
   B component, 500 ml
   Vitamin B_1, 1 mg/ml, 10 ml
   YT medium, 10 ml (the YT medium increases mating efficiency at least 2-fold)
   20% sugar, 12.5 ml

PROCEDURE

Grow the cells overnight in tubes containing 1 ml of the growth mix. In the morning add 3.5 ml of the mating mix to each tube. This is conveniently done with a Repipet (Labindustries, Berkeley, California). Grow the cells to a density of 1–2 × 10^8/ml. This takes 2.5–3 h. Using a 1-ml pipette, place one-drop quantities of the cells (approximately 0.05 ml) in the appropriate wells of the mating blocks. It is convenient to use the lid as a prop and marker. Incubate the block at 35 °C for 1.5–3 h. Spot onto the trays of selective medium with the transfer prongs. The prongs are dipped in the wells and touched to the agar surface in the trays. When seeking rare recombinants, it is helpful to spot 2 or 3 times to increase the number of cells spotted. The prongs are sterilized only when many different wells are to be spotted. To sterilize, flush with hot running water, dip in ethanol for 15 sec, invert, and blow dry with a hair dryer. The hot water flush warms the prongs enough to assist the evaporation of the ethanol, and the hot air flow from the dryer finishes the drying operation without overheating. The sterile prongs require no additional cooling before the next transfer.

USING TRANSPOSONS IN STRAIN CONSTRUCTION

Most of the genetic operations described thus far are straightforward, convenient, and practical for occasional strain constructions. However, for extensive strain construction or for heavy genetics, better although somewhat more finicky methods exist. Thus, it is beneficial to be aware of the wealth of genetic information contained in *Experiments in Molecular Genetics* by Miller (1972), as well as the genetic tricks and versatility of drug resistance factors described in "Genetic engineering in vivo using translocatable drug-resistance elements," by Kleckner et al. (1977).

Chapter 2

Bacteriophage Lambda

Over the years bacteriophage lambda has been intensively studied because it is fairly small, can integrate and excise from the host chromosome, and has a sophisticated regulatory system (Ptashne et al., 1976). In its own right, lambda phage continues to be important in molecular biology; however, its importance has been greatly increased through its use as a vector in genetic engineering (Thomas et al., 1974). For many engineering projects it is the best vector for the first steps in cloning genes from higher organisms. This is largely due to the high transformation efficiency that can result from the in vitro packaging of recombinant DNA into the phage heads of lambda (Kaiser and Masuda, 1973).

In this chapter we describe some of the basic manipulations of lambda phage. These manipulations are useful either in the study of lambda or in the study of genes cloned into lambda. In addition to the basic methods described in this chapter, in Chapter 6 we will describe other methods that are of particular use in genetic engineering.

TWO USEFUL MUTANTS

CI_{857}

Ordinarily, lysogens of lambda are induced 1–2 h after the DNA of the host has received substantial damage by, for example, UV irradiation. (For an overview, see *Molecular Biology of Bacterial Viruses,* Stent, 1963.) Since UV and chemical treatments are inconvenient methods of induction, a mutation has been isolated that makes the lambda repressor heat labile. This mutation, CI_{857} (Sussman and Jacob, 1962), has been so useful that it is now rare to find a lambda phage without it. Lambda phage carrying the mutation induce at temperatures a little above 37 °C. Typically, when lysogens are being constructed or when it is necessary to prevent induction of the phage, cells are grown at 35 °C. The phage is induced by growing cells at 42 °C for more than 5 min.

S_7

Ordinarily, growth of lambda ceases and the host lyses 45–90 min after the phage begins growth (see *The Bacteriophage Lambda,* Hershey, 1971). A typical burst size is about 100 phage particles. However, lysis is prevented and synthesis of the phage is not shut down if the phage contains a mutation in the S gene (Adhya et al., 1971). Such a mutation permits the phage to grow in cells for up to 5 h and yields up to 1000 phage particles per cell. More importantly, since the cells do not lyse, they may be concentrated by centrifugation before artificial lysis. Also, the low speed centrifugation necessary to pellet bacterial cells is much easier to perform than the high speed centrifugation necessary to pellet phage particles.

Figure 2.1. The appearance of normal *Escherichia coli* and the appearance of *E. coli* in which lambda S_7 mutants have been induced and permitted to grow for approximately 3 h.

A conditional mutation, S_7, has been isolated to assist in working with S mutations (Goldberg and Howe, 1969). This mutation permits the convenient titering of the phage because it is a nonsense mutation that can be suppressed by Su_{III}^+ hosts to yield an active S gene protein. Cells in which S^- mutants have grown for more than 2 h are more elongated than normal (Figure 2.1). This permits rapid verification of the effectiveness of infection or induction.

TITERING

Problems are rarely encountered when titering lambda because most strains and most varieties of plates are satisfactory. Occasionally, however, one forgets to use the proper plating strain, for example, not using an Su_{III}^+ strain when titering S_7 mutants.

PROCEDURE

1. Grow the suitable plating bacteria overnight in 5 ml of YT medium.
2. Pellet the cells by centrifugation at 5000 $\times$ *g* for 5 min. It

is convenient to use the heavy-wall Sorvall centrifuge tubes for this.

3. Resuspend the cells in 5 ml of 0.01 M $MgSO_4$. These cells will keep for a week in the refrigerator.
4. To titer, dilute phage serially through tubes containing 5 ml of YT medium.
5. Add 0.1 ml of plating cells to 13 × 100 mm sterile tubes and add 0.1 ml of the appropriately diluted phage.
6. Incubate the cells and phage at 37 °C for 10 min to permit adsorption.
7. Add 2.5 ml of TB medium-top agar at 48 °C, roll the tube briefly between the palms of your hands to mix without producing bubbles, and immediately pour onto a plate. (Agar solidified in a pipette hinders later cleaning. To prevent this, immediately after pipetting the agar, draw about 5 ml of water up into the pipette and then blow it out.) One common problem in pouring the top agar is that it begins to harden too early. Since the 2.5 ml does not flow to cover the entire surface of the plate it is necessary to shake the plate a little with a circular motion in order to spread the agar. Usually, there is insufficient time to put down the tube and its cap before shaking the plate. With about a minute's practice you can learn how to hold the lid of the tube in one hand, take the lid of the plate off with the same hand, and then pour the agar and shake the plate with the other hand or with both hands. When this is done, replace the lid of the petri plate and drop the tube and its lid into a dish pan of water.
8. By estimating the position of 12.5 ml in a 10-ml pipette, you can pipette agar into 5 tubes in one pass. Speed permits all of them to be mixed and poured before any have begun to solidify.
9. If the plates are too dry, the plaques will be tiny and hard to see. If the plates are too wet, they will perspire and produce blotches of infected cells. Plates at the optimum dryness do not have condensation on the lids but the agar surface has not yet wrinkled.

GROWING PLATE STOCKS

Plate stocks provide the simplest method for amplifying a phage from a single particle to the number of phage needed in most experiments. Of particular convenience is the fact that the resultant phage stocks have a high titer, 10^{10}–10^{11}/ml. This results from the fact that the host bacteria grow on the surface of the plate, but obtain nutrients from the entire 25 ml of medium in the plate. The resultant phage can be harvested in just enough liquid to wet the surface, 2.5 to 5 ml. Unfortunately, this surface-area-to-volume scaling cannot be used in all situations, and when substantially greater amounts of phage are required, the cells are most conveniently grown in the medium rather than on it.

It is most important that very fresh moist plates be used when growing lambda stocks. If a plate is more than 2 days old, it will still work well if 1 or 2 ml of broth or water are added about 12 h before the plate is used. If a plate is more than 3 or 4 days old, it may work poorly even with this additional liquid. Thick, 20 mm-deep plates work best. Use fresh bacteria which have grown to $1-2 \times 10^9$ cells/ml in YT medium. Cultures which have grown overnight to stationary phase work well. Mix bacteria, 0.05 to 0.1 ml for each plate, with less than 0.5 ml of phage and adsorb for 15 min at 37 °C. Unlike titering experiments, it is generally not necessary to include Mg^{2+} in the medium. However, in making a plate stock, either adding Mg^{2+} to 0.01 M or resuspending bacteria in 0.01 M $MgSO_4$ probably does not hurt. After adsorption, add 2.5 ml of melted TB-top agar at 48 °C and pour the mixture on a plate. The optimum phage input depends on the plaque size (burst size) of the particular phage. For wild-type phage, 10^4-10^5 particles per plate is good. With mutants that yield small plaques, 10^6 particles per plate may be necessary. Clear phage produce higher titers; therefore, if possible, nonsense clear mutants should be grown on Su^- hosts and CI_{857} mutants (tsCI) at > 37 °C.

When initially using a stock that has been stored for some time, a new stock obtained from someone else, or a new mutant, it is wise to purify the phage by "cloning" (old definition), that is, by growing a new stock from a single plaque. This insures that no rare, but more rapidly growing mutants in the original population will eventually outgrow the desired phage. Single plaques can be isolated either by the streaking method described below or by plating out. An isolated plaque can be picked by using the tip of a sterile Pasteur pipette to extract a plug of agar containing the chosen plaque. Vortex the plug with 0.5 ml of lambda dilution buffer and a drop of chloroform. After 30 min at room temperature sufficient phage will have been extracted so that 0.1 ml of the suspension, free of chloroform, can be used to make a plate stock. Such an extraction of phage from an isolated plaque requires that the plaque be not more than 12 h old. Thereafter adsorption of phage to bacterial debris will reduce the concentration of free phage. Similarly, phage will readsorb to bacterial debris during the extraction if the suspension is left too long at room temperature. If the suspension is not going to be used within an hour, it can be refrigerated overnight with little loss of free phage.

It is sometimes difficult to determine the correct incubation time when making plate stocks, but this is fairly important. Ideally, plaques on the phage plates should be just approaching confluence at the time of harvest. In order not to incubate too long, it is helpful to incubate an additional plate that contains cells but no phage. When the plates are ready, the bacterial growth on this control plate should be noticeably turbid. Usually 5.5 h at 37 °C is satisfactory, although small plaque formers often take longer to give maximum yield. As described above, leaving the plates too long produces a substantial decrease in yield.

To harvest the phage, scrape the top agar into a sterile centrifuge bottle with an L-shaped Pasteur pipette. If the plates are as fresh as they should be, the top agar will come off very easily. Add an equal volume of YT medium or phage buffer and then add 0.25 ml $CHCl_3$. Use polyethylene or polypropylene centrifuge bottles instead of the clear polycarbonate bottles since polycarbonate is more soluble in $CHCl_3$. Shake thoroughly and then pellet the agar and cell debris by centrifugation at about 6000 $\times$ *g* for 10 min. The Sorvall centrifuge with the SS34 rotor, 10,000 rpm for 10 min, works well. Reextracting the agar pellet gives more phage.

Titers will usually fall between 10^{10} and 10^{11} particles per ml in a total volume of about 4 ml. If necessary, the stock can be concentrated by a 90-min spin at 22,000 rpm in the Beckman angle 30 rotor, 40,000 $\times$ *g*. Discard the supernatant and resuspend each pellet in 0.5 to 1 ml of phage dilution buffer by gently shaking overnight at 4 °C. Rapid mechanical resuspension kills most of the phage, possibly by breaking off their tails. Spin out the remaining agar at 6000 $\times$ *g* for 5 min. Concentration by this method usually kills some of the phage.

COMMENTS

1. From a single phage plate one can usually obtain about 5 $\times$ 10^{10} phage particles. When greater numbers are required, it is most convenient to use up to about ten plates. For still greater numbers of phage, it sometimes is most convenient to use larger petri plates. They come in sizes up to medium-sized pizzas. Pyrex baking trays are a good alternative. When airborne phage contamination is a problem in a laboratory these schemes of multiple plates increase chances of disaster. The larger plates or trays seem to have a greater propensity for condensation problems, so it often helps to permit the agar medium to solidify with the lid off.
2. These methods apply with almost no modification to other members of the clan of lambdoid phage: ϕ80, ϕ81, ϕ82, ϕ21, ϕ381, ϕ424, and ϕ434. It has been found that the tail fibers of ϕ80 are temperature sensitive, so this phage will not form plaques at 42 °C. One of the most popular hosts for growing lambda, the C600 strain of *E. coli,* is resistant to phage T1, which plagues some laboratories. This bacterial strain is also resistant to ϕ80, since these two phage use the same receptor on the cell surface.

LARGE-SCALE GROWTH IN LIQUID

When it is necessary to isolate more than 10^{12} phage it is easier to do so by growing the bacteria and phage in liquid rather than on an agar surface. If you are growing phage in liquid, it is better to cross the CI_{857} and S_7 mutations (see p. 27) onto the phage and to make a lysogen of the phage. Finally, you should make that lysogen lambda-resistant. If this is done, the phage can be induced rather than grown by infection, up to 1000 phage will be made per cell, the phage can be easily harvested, and the yield will be larger because the phage will not adsorb to bacterial debris in the lysate. However, sometimes a lysogen cannot be made and infection is the only course available.

One complication to infecting cells and growing them in liquid is that the ultimate yield of phage is strongly dependent upon the initial infection multiplicity. A possible explanation for this is that infected cells subsequently lose their ability to adsorb. If this is the case, then the fraction of uninfected cells present at the time of phage release will drastically affect the amount of phage lost by adsorption to debris. This fraction can be large even though most cells have been infected because the uninfected cells will continue to divide until the phage are harvested. For this reason, infect with a multiplicity of at least 6 when cells are to be grown several hours after infection and when they will be concentrated before lysis, that is, when using S gene mutants.

PROCEDURE USING INFECTION

1. The procedure works when scaled up or when scaled down to as little as 20 ml.
2. Grow cells to 1×10^8 per ml in 1.3$\times$ concentrated YT medium at 35 °C and then add maltose to a concentration of 0.13%.
3. When cells reach 2×10^8 per ml, add $MgSO_4$ to 0.007 M and phage CI_{857}, S_7 to a multiplicity of at least 6.
4. After a 15-min adsorption, insure induction of phage by raising the temperature to 42 °C for at least 5 min. Then lower the temperature to 35 °C for the remaining 3.5 h of growth.
5. Spin down the cells. For large volumes, the culture can be passed through a DeLaval Gyro Tester (p. 13) at a rate of 500 ml/min. Otherwise, spin down at 5000 rpm in a Sorvall GSA or SS34 for 5 min, 5000 $\times$ *g*.
6. Resuspend 50 g of cells in 150 ml of lambda dilution buffer.
7. Add 1 mg of crude pancreatic DNase I and 2.5 ml of chloroform per 200 ml of resuspended cells. When smaller quantities of cells are being processed, it is convenient to dissolve DNase in 10^{-4} M HCl and to add the appropriate volume to the cells. The DNase remains active for several months when stored in the refrigerator in this HCl solution.
8. Shake the mixture at 35 °C for 30 min.
9. Remove bacterial debris by two successive 20-min centrifugations at 8000 $\times$ *g*.
10. Note that the amount of DNase necessary to keep the lysate from becoming a gelatinous mess is not critical,

but since the usual product desired is phage DNA it is good to minimize the DNase. It is necessary to add only sufficient DNase so that the glop can be managed with a pipette.

MODIFICATIONS FOR HEAT INDUCTION

1. Be sure the cells are lambda-resistant and that the bacterial stock contains the phage you want.
2. Grow cells in YT medium or 1.3 times concentrated YT medium at 35 °C to 2×10^8 cells/ml and heat the culture to 42 °C for at least 5 min. When inducing a culture in a flask this is easily done by immersing an ethanol-wiped thermometer and shaking the flask in the flame of a Bunsen burner. Twenty minutes are usually required to bring a carboy to 42 °C (p. 11).
3. Phage growth proceeds at 42 °C but yields are higher if the temperature is lowered to 35 °C after induction. Generally harvesting 3.5 h after induction maximizes phage yield.

PURIFICATION

Purification of lambda by centrifugation through CsCl provides a convenient method for removing bacterial DNA and extraneous bacterial debris. Classically this step was performed as an equilibrium density gradient centrifugation; however, this is a fairly time-consuming procedure and can be replaced by two shorter centrifugation steps. In the first step the lysate is centrifuged down through a layer of CsCl of density 1.3. The phage pass through this layer into a layer of CsCl of density 1.5. All low density contaminants and most other small contaminants stay above the CsCl layers. In the second centrifugation step, the phage are floated up through a layer of CsCl of density 1.7. This second centrifugation removes all high density contaminants such as bacterial DNA. The two centrifugation steps can be performed in 6 h.

Stock CsCl solutions are made up as follows:

Density	CsCl (g)	Phage dilution buffer (ml)	Index of refraction
1.3	10	22.3	1.362 ± 0.005
1.5	10	12.6	1.381 ± 0.005
1.7	10	7.9	1.392 ± 0.005
1.9	a saturated solution of CsCl in phage dilution buffer		

The index of refraction is read on a Bausch & Lomb refractometer. A solution of 20% sucrose in phage dilution buffer should also be made.

The first centrifugation is on a CsCl block gradient pre-

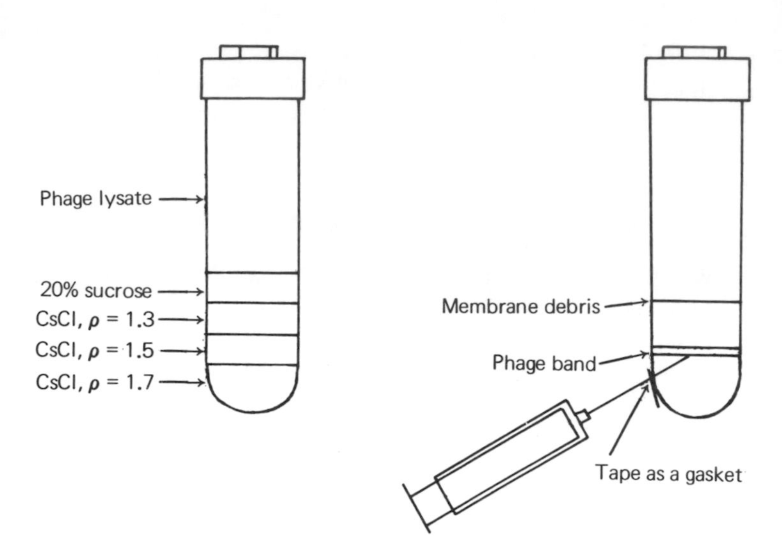

Figure 2.2. The appearance of a filled ultracentrifuge tube before centrifugation showing the 3 layers of CsCl, the layer of 20% sucrose, and the phage lysate, which may be overlaid with mineral oil. On the right is shown a centrifuge tube following centrifugation in which the whitish membrane debris is usually visible as well as the more bluish phage band. The hypodermic syringe is approaching the phage band from below, having penetrated the centrifuge tube through a layer of tape which acts as a gasket to reduce leakage.

pared as shown in Figure 2.2. After the first layer is in the tube, the successive layers should be pipetted in very slowly, holding the tip of the pipette on the edge of the tube just above the liquid surface. During addition of the CsCl and sucrose layers, there will be some mixing between the layers. However, unless you have been violent, there will be no problem and four discrete layers will be visible when you are finished. After the phage lysate has been layered, fill the tube, if necessary, by the addition of mineral oil. Then centrifuge the tube at 22,000 rpm, 50,000 × *g*, for 90 min in a Beckman SW27 or SW40 rotor at 4 °C and allow it to come to rest either with or without using the centrifuge brake.

After centrifugation, the phage will form a bluish white band near the bottom of the tube. This can be seen more easily when a flashlight is shone obliquely on the tube. The phage band should not be confused with the other white band or bands of membrane and cell wall fragments which will be seen at the 20% sucrose level of the tube. Extract the phage band with a 5-ml hypodermic syringe fitted with a 22-gauge needle. Clean the outside of the tube with tissue paper and stick a small piece of paper adhesive tape to the tube about 1 cm below the phage band. Insert the needle through the tape and into the tube so that the needle enters the tube about 1 cm below the band and is inclined upward toward it. Push the needle up to the bottom of the band and slowly extract the phage band. Note that if you are to avoid contaminants you will not be able to withdraw all of the phage.

For the second and upward centrifugation step, mix the phage solution extracted from the first centrifugation with an equal volume of saturated CsCl solution. Put this in the bottom of a centrifuge tube and layer CsCl of densities 1.6 and 1.4 above it. Make these by mixing equal volumes of CsCl of densities 1.7 and 1.5 and 1.5 and 1.3, respectively. Fill the remaining volume of the tube with mineral oil and centrifuge at 22,000 rpm for 90 min at 1 °C, as before. Although angle rotors may be used for these steps, it is more convenient to use a swinging bucket rotor. At the end of the centrifugation collect the phage band as before. Lambda phage seem to be most stable when stored in the CsCl solution at 4 °C.

GENETIC CROSSES

The need for performing a recombination with lambda phage arises when it is necessary to separate or join mutations on the same phage. The recombination is performed by infecting host cells with each of the parental phage at such a multiplicity that each cell is likely to be simultaneously infected by both of the input types of phage. Growth of the two types of phage in the cell permits both host and phage-specified recombination mechanisms to produce recombinants. Phage resulting from such an infection are then plated out and individual plaques are screened to identify the desired recombinant. Considerable data have been obtained in recent years about this process, but information sufficient for most purposes is contained in a review on general recombination (Signer, 1971). One of the main objectives in designing experiments in which recombinants will be sought is to construct them so that the desired recombinants can easily be identified. A variety of tricks have been used to make this easier, and since most depend on the particular mutants being used, it is not practical to list them. However, it often makes sense to use input phage of different immunity so that one of the parental phage can be immediately identified. Often the other parental phage can be arranged to have a nonsense mutation which permits it to be easily identified by spotting onto Su^- and Su^+ hosts. Host range mutants can also be used to mark input phage. Occasionally, when rare mutants are being sought, it can be arranged that the input phage have different densities from the desired recombinant and the recombinant can be found in the appropriate position on a CsCl equilibrium gradient.

PROCEDURE

1. Grow the C600 strain overnight in 5 ml of YT medium containing 0.4% maltose. As mentioned before, this particular strain is T1-resistant and also carries the Su_{II} suppressor. Maltose is placed in the medium to induce greater synthesis of lambda receptor sites. Glucose will depress the receptor synthesis.
2. Dilute the saturated overnight culture 100-fold into fresh medium and grow to a density of 2–4 $\times$ 10^8 cells/ml.
3. Spin the cells down (10 min at 6000 $\times$ *g*) and resuspend in 10^{-2} M $MgSO_4$ at about 10^9 cells/ml.
4. Mix 0.1 ml of each phage at 5 $\times$ 10^9 particles/ml and

then add 10^8 bacteria. Depending on the cross, you may want to vary the multiplicity.

5. Let the phage adsorb for 15 min at 37 °C.
6. Add 5 ml of cold 10^{-2} M $MgSO_4$. It may be necessary to stimulate rare recombination events by irradiating phage DNA in the cells. A 10-sec exposure at 40 cm from two 15-W germicidal lamps that are backed by a reflector, 220 erg/mm^2, works well. Pour the cell suspension into a small petri dish which is sterile and chilled on ice. Gently swirl the dish while the UV lamp is on.
7. Pour the cells into 5 ml of YT medium or dilute 10- or 100-fold into YT medium, depending on the concentration of phage which you will plate. In general, it is best to dilute 100-fold so as to prevent readsorption after cell lysis and to avoid the necessity of making many dilutions later.
8. Grow for 1.5 h at 37 °C. Add 2 drops $CHCl_3$ and vortex. Chill on ice for at least 15 min to increase cell lysis and vortex again.
9. Plate or do whatever else is required to select the mutant. Usually you plate to obtain the titer and to obtain isolated plaques. It is frequently best to plate on a nonlysogenic indicator since the plaques of a phage plated on a lysogen usually contain some phage from the lysogen. To test for the required recombinant, spot the plaques from the nonlysogenic indicator onto various TB plates that have the desired indicators in their soft agar overlayer.
10. The recombination frequency between genes CI and R is about 5% if cells have been given UV treatment.

SCORING PLAQUES

Frequently it is necessary to score plaques in order to identify a particular new mutant, to identify a desired recombinant, or to confirm the identity of a strain. Usually this is done by toothpicking candidate plaques onto several different lawns of bacteria. The ability of lambda to grow on these different lawns or to form a particular type of plaque can then be determined.

PROCEDURE

1. Toothpick plaques onto the restrictive lawn first. For example, first stab onto an Su^- lawn, then onto an Su^+ keeper lawn, and finally onto an Su^+ lysogen.
2. In general, don't pick plaques that have been plated on a lysogen. There are almost always enough prophage present to confuse scoring.
3. It is helpful to dip the tip of the toothpick into a sterile dye such as eosin yellow or methylene blue before stab-

bing onto the lawns. This will keep track of the positions on the tester plates and also indicate which plaques have already been tested.

MAKING STRAINS LAMBDA-RESISTANT

When making large-scale preparations of lambda phage particles by inducing a lambda lysogen it is essential that the lysogen be resistant to phage infection. Otherwise, the phage will adsorb to the bacterial debris and the yield can be poor.

The basis of this method for selecting lambda-resistant mutants is that lambda adsorbs to cells via a protein in the outer membrane that is necessary for the uptake of maltose. Frequently, then, lambda-resistant mutants have inactivated this protein or lost the ability to synthesize it, therefore becoming Mal$^-$. In this method for the selection of lambda resistance, cells are infected with lambda and then plated out on maltose-indicating plates to determine which of the surviving cells are Mal$^+$ and which are Mal$^-$. Since infection with wild-type lambda could yield a sizeable fraction of cells surviving as lysogens, a virulent lambda, one which can grow even in the presence of lambda repressor, is used for the selection. Hence, lambda-resistant mutants will be an appreciable fraction of the Mal$^-$ survivors. Presumably this method is not useful for the isolation of cells resistant to the other lambdoid phages since these phages probably use other cellular structures for adsorption.

PROCEDURE

Grow an overnight culture of the desired strain in YT medium containing 0.2% maltose. Mix 0.1 ml of this culture with 0.1 ml of a virulent lambda phage, λ*vir,* so that there are about 10 phage per bacterium. Allow the phage to adsorb 10 min at 37 °C and plate 0.1 ml of the mixture and 0.1 ml of a 10-fold dilution onto EMB-maltose (1% maltose) plates. Survivors will generally be of 3 types:

1. Mucoid colonies. Discard these.
2. Mal$^+$ (red) colonies. These are often resistant to lambda, but sensitive to the lambda host range mutant, λ*h*. You probably want the colonies resistant to all lambdas.
3. Mal$^-$ (pink) colonies. These are usually resistant to both lambda and λ*h*. (For an explanation, see Schwartz, 1967.) Purify the pink colonies and then test them by streaking across a line of λ*vir* that was formed by streaking the phage on a YT plate at a concentration of 10^8–10^9 particles/ml. Note that Mal$^-$ lambda-resistant strains can usually be made lambda-sensitive by selecting for Mal$^+$.

TESTING COLONIES FOR THE ABILITY TO GROW LAMBDA

Often it is necessary to determine whether or not a particular bacterial strain can grow lambda or a particular lambda mutant. A cell's inability to grow lambda could arise either by mutation or by being immune due to the presence of a prophage.

Cells are spread on an agar surface which has been previously spread with lambda phage at a density such that most individual cells will not initially be in contact with a phage. Then, as the bacterial cells divide and grow into colonies, they come into contact with one or more phage. These phage begin to attack the colonies and can take appreciable "bites." Usually, sufficient uninfected cells remain so that survivors can be purified for subsequent use. "Bitten" and "unbitten" colonies are clearly distinguishable.

The phage stock must be about 1×10^{10} particles/ml, and revertants in the stock must be less than 5×10^{3}/ml. Spread 0.1 ml of this phage stock onto a TB, YT, or EMBO plate containing no added sugar. After the plate is dry smear the candidate bacterial strains ($\sim$ 20 per plate) on different portions of the surface with a toothpick and then grow them overnight.

In general, a colony able to support phage growth will have ragged edges whereas a colony unable to support phage growth will have a smoother appearance. On EMBO, colonies that support phage growth will also turn dark.

MAKING LYSOGENS

Frequently it is useful to lysogenize a bacterial strain with a lambda phage. It might be necessary to have a particular lysogen to use in a lawn when toothpicking for particular mutants or recombinants. It might also be necessary to use a lysogen to obtain large quantities of phage (p. 31). Occasionally, it is more convenient to store a particular lambda mutant as a lysogen rather than as a phage stock.

PROCEDURES FOR CONSTRUCTION

1. It is easiest to lysogenize wild-type lambda. Simply plate the phage so as to obtain several hundred plaques per plate and then purify lysogens by streaking out cells that have survived at the center of a turbid plaque. Test the clones for lysogeny as described below.
2. In general, the following is the most useful method for non-wild-type lambda. On a YT plate use a loop or a 0.1-ml pipette to make a line of a bacterial culture that has been grown overnight in TB plus 0.2% maltose. After the bacteria have dried, spot the phage at a concentration of

about 10^9 particles/ml onto the bacterial line and then incubate at 32 °C. It is advisable to test for bacteria in the phage stock by spotting the stock on the same plate, but well away from the bacterial line. The next day, purify single clones from areas where the phage were spotted. The following day, test single clones as described below. This method can often be used when the bacterial host will not allow the phage to make plaques.

3 *Sus* phage (nonsense mutants of lambda phage are called Sus) can be lysogenized in Su^- hosts by infection in liquid of high multiplicity. Adsorb phage to bacteria that have been grown overnight in TB medium or YT plus 2% maltose. This adsorption is done at a multiplicity of 5–20 for 20 min at room temperature. Most of the bacteria will survive, so dilute the adsorption mixture accordingly and plate on TB plates. If less than one in 20 of the tested clones are lysogens, something is probably wrong with the phage, the host, or the method used.

4. Some lambda strains such as N^- and Int^- must have helper phages in order to form lysogens. When using a helper phage to make a lysogen, it is possible that the two phages will recombine so that the resident prophage will not be the desired type. Although the frequency of this event is low compared to the frequency of integration, it often pays to determine whether the prophage is of the desired genotype.
   a. N^- phage do not express the lambda integration function (Signer, 1970). Consequently, to make an Su^- (λN^-) lysogen, one must use a helper phage which supplies the N protein. Infect the host with the N^- phage and with a helper, like $\lambda imm^{434}b104$. Use at a multiplicty of five of each. A phage like *b*104 lacks the lambda attachment site, so that most of the lysogens obtained will be single lysogens of the N^- phage. Test the putative lysogens for immunity to both lambda and 434.
   b. Int^- phage lack functional Int protein so an Int^+ helper should be used. Coinfect with the two phage at a multiplicity of five of each to make a lysogen. To determine whether the helper phage has integrated along with the one desired, you must choose the helper with discrimination. For example, use λimm^{434}, and subsequently test the lysogens for 434 immunity. Alternatively, use an Int^- phage such as CI_{857}. The desired lysogen will die at 42 °C.

COMMENTS

1. $\lambda b2$ cannot integrate efficiently into the wild-type bacterial attachment site, but it can integrate efficiently into combination phage-bacterial attachment sites (see Hershey, 1971, Chap. 6).
2. Double lysogens: lambda lysogenizes a lambda lysogen

very poorly since *int* is not expressed. Double lysogens can be constructed easily if the second phage has a different immunity, thus allowing the second phage to express *int.* These double lysogens segregate out one of the prophages by a recombinational process. Consequently, if one wants a stable double lysogen, it is best to begin with a $RecA^-$ host.

3. For some of the methods described above, you can select for the lambda lysogens by plating the infected bacteria on a plate spread with about 10^9 CI phages. You must, however, be aware of the possibility that this "challenge" phage may also become a prophage.

TESTING LYSOGEN CANDIDATES

1. The best method is to test for immunity by using a toothpick to streak the candidates across dried streaks of tester phage on an EMB-maltose (0.1% maltose) plate. Two vertical lines of phage (for example, λCI and λ*vir* [Ptashne, 1971]) are first put down with a 0.1-ml pipette. Stab the toothpick into the candidate colony and then onto a keeper plate, and finally drag it lightly first across the phage whose immunity is to be tested and then across the λ*vir.* Lysogens should be immune to the tester phage, but sensitive to λ*vir.* The λ*vir* streak is a control to insure that the bacterium is not simply resistant to infection. One should purify the lysogen at least twice before using it.
2. To identify CI_{857} lysogens, one can usually test whether the lysogen is unable to grow at 42 °C.
3. Sometimes it is easiest to test for the prophage by testing for phage release. First stab the clone onto a keeper plate and then onto a lawn of permissive bacteria. Prophage in the latter plate can be induced by UV or heat.

STREAKING FOR SINGLE PLAQUES

Phage can be purified in much the same way as bacteria, i.e., by streaking across a plate to dilute and obtain a single isolate. To do this with phage, streak a loopful of phage suspension on a lawn of indicator bacteria. Streaking different phage on the same plate is also a useful method for comparing plaque size.

A necessary tool is a light-gauge platinum wire, 0.02–0.03 mm in diameter with a 2-mm loop at the end.

1. Add 2.5 ml of top agar to 0.1 ml of indicator bacteria and pour on a phage plate. Then allow the plate to sit at room temperature for at least 30 min before streaking. This wait permits the lawn to harden adequately.
2. Use a loopful of the phage suspension to make a zigzag streak on a sector of the plate. It is difficult to obtain iso-

lated plaques by this technique when the phage suspension is above 10^8 particles/ml.

SELECTING DELETIONS

The great utility of deletions in the analysis of gene function recommends their wide usage. Although deletions may be made in DNA cloned in plasmids by a straightforward use of restriction and ligation enzymes, in some situations it is more efficient to use in vivo methods to isolate deletions of lambda genes or of foreign DNA inserted into lambda. The principles of deletion isolation in lambda are fully described by Parkinson and Huskey (1971) and Davis and Parkinson (1971). Apparently, removing Mg^{2+} from phage causes most of the phage particles to pop. This tendency to pop is increased by heating the phage. Any phage particles which possess less than normal quantities of DNA in their heads are less likely to pop and so deletions are selective survivors. The same principles apply to liquid or plate growth of phage. The best method for lowering Mg^{2+} in a phage lysate is to add a chelator such as EDTA.

This approach can also be used during the growth of plate stocks. However, the use of EDTA to reduce the Mg^{2+} concentration is tricky with plate stocks because the EDTA concentration must be precisely controlled in order to leave just a trace of free Mg^{2+} for the cells. A better method of lowering the free Mg^{2+} on plates is to use pyrophosphate as a chelator. Pyrophosphate concentration appears not to require such careful adjustment.

Two additional factors are important in isolating deletions with EDTA or pyrophosphate. First, if the selection is performed in liquid, about 1% of the phage particles survive the chelation-heating step despite the fact that they lack a deletion. These anomalous survivors are thought to have an abnormal coat, for on regrowth they acquire normal EDTA sensitivity. Thus, for complete selection of deletions in liquid it is necessary either to use several cycles of killing and regrowth or to follow a simple treatment in liquid with some other selective step. The straightforward way to accomplish multiple cycles of killing and regrowth is first to perform a chelation-heating step in liquid and then to plate the survivors on a pyrophosphate plate. If this procedure is used virtually every plaque derives from a deletion phage. The exact details of this approach as used in one application are described by Schleif and Lis (1975).

The second factor is that the survival probability of the deletion phage should be as high as possible relative to the survival probability of the parental phage. For input phage with DNA substantially shorter than that of wild-type lambda, less than, for example, 95% of the length, the differential survival probability between the input and deletions of the input becomes markedly lower. Hence, it is useful to begin with a parent DNA that is as large as possible.

Chapter 3

Enzyme Assays

Enzyme assays are essential to molecular biology and often many types are necessary for the completion of even the most simple experiments. There are thousands of known enzymes and usually several ways to assay any one of them. We will discuss just a few enzyme assays, those which we have used so frequently that most potential difficulties have been anticipated, and those which possess features which can be incorporated into assays for a wide variety of other enzymes.

β-GALACTOSIDASE

The ease in assaying β-galactosidase has led to its nearly ubiquitous use in molecular biology. Most recently this assay has found important use as an assay for protein and gene fusions (Casadaban and Cohen, 1979; Casadaban et al., 1980), as well as in determining progress in sophisticated recombinant DNA constructions (Guarente et al., 1980). These are often planned so that β-galactosidase will be synthesized when fusions, deletions, or other DNA rearrangements have occurred.

This enzyme is conveniently assayed by measuring its cleavage of the colorless ONPG (orthonitrophenyl-galactoside) into galactose and a yellow compound, nitrophenyl. A simple colorimetric assay for nitrophenyl permits accurate quantitation of the amount of enzyme present. Furthermore, since the product of the enzyme reaction is visible, a problem inherent in most other enzyme assays is easily avoided. Thus, the visible reaction product allows you to stop the assay reaction when sufficient product has been generated. In most assays this cannot be done, and you must either perform initial assays to determine a sample's approximate activity and then perform the definitive assays at the correct enzyme dilution, or you must prepare multiple dilutions of the sample so that one dilution will be within the assay's range of sensitivity and linearity.

The β-galactosidase enzyme has several other useful features. It is active in a wide variety of buffers and may even be assayed in whole cells if they have first been made permeable to ONPG by vortexing in toluene. Also, the turnover number of the enzyme is known, so the actual number of enzyme molecules present in an assay may be determined if the assay is performed in the buffer described below. Other enzyme-substrate combinations, e.g., alkaline phosphatase and *p*-nitrophenylphosphate, will also synthesize nitrophenyl, although, in this case, *p*-nitrophenyl. The principle of their assay is thus very similar.

PROCEDURE AND COMMENTS

1. β-Galactosidase synthesis may be induced by growing *E. coli* in the presence of IPTG (isopropylthiogalactoside) or in a medium that has lactose as the carbon source. IPTG at 2×10^{-4} M gives half-maximal induction of the *lac* operon (Gilbert and Müiler-Hill, 1966). The presence of glucose in the medium inhibits inducibility of the *lac* operon (Epstein et al., 1975) and usually glycerol or succinate is provided as a carbon source when IPTG is used to induce. The glucose-induced catabolite repression can be overcome by including 10^{-3} M 3′,5′-cyclic AMP in the growth medium (de Crombrugghe et al., 1969).
2. If whole cells are being assayed, make them permeable by vortexing for 15 s with 1/20 vol. toluene. Cells in culture medium can be used or cells can be concentrated by centrifugation and resuspended at less than 10^{11} cells/ml in assay buffer (0.1 M sodium phosphate, pH 7.0; 0.001 M $MgSO_4$; 0.1 M 2-mercaptoethanol) before vortexing with toluene.
3. If the volume of the sample to be assayed is less than 0.8 ml, use assay buffer to dilute it to 0.8 ml.
4. Warm the reaction tubes containing 0.8–1.0 ml of cell suspension or of more purified enzyme to 30 °C. Add 0.2 ml ONPG (4.0 mg/ml in assay buffer) and shake.
5. Incubate until a pale yellow color develops. This incubation can be as long as 30 h. Then add 0.5 ml 1 M Na^+-carbonate to stop the reaction and finally read the absorbance at 420 nm. Include suitable controls for the spontaneous hydrolysis of ONPG as well as light scattering from whole cells.
6. Occasionally, a precipitate forms when the reaction mixture is poured into a cuvette. This often due to toluene remaining in the mixture after step 2. The toluene can be removed by placing the tubes in a desiccator and evacuating for 30 min with a water aspirator.
7. Be aware that the definitions of units vary. However, Craven et al. (1965) have measured the pure enzyme's turnover number in the buffer described above, so their units

are frequently the most useful. Utilization of their units and several other facts about this enzyme yields the following handy list of information.

a. One unit of enzyme hydrolyzes 10^{-9} moles ONPG/min in the assay buffer described above.
b. Pure enzyme is 340,000 units/mg.
c. Monomer molecular weight of β-galactosidase is 130,000.
d. Molar extinction coefficient of nitrophenyl is 4860 at pH 10 (Doub and Vandenbelt, 1949).
e. Enzyme sufficient to produce 1 $OD^{1\ cm}_{420\ nm}$/min is 4.45×10^{12} monomers.
f. A typical uninduced level of β-galactosidase is 5 monomers per *E. coli* cell and a fully induced level is 4000 monomers per cell.

RNA POLYMERASE

E. coli RNA polymerase is widely used to transcribe DNA into RNA. Here we give a simple assay for *E. coli* RNA polymerase activity developed from that described by Burgess (1969). When RNA polymerases from other organisms are being assayed, details of the assay will differ somewhat, but the principles will remain the same. This assay measures the synthesis of high molecular weight RNA from the precursor mononucleoside triphosphates. Radiolabelled precursor is used and after the reaction the precursor is separated from the polymerized product by trichloroacetic acid precipitation. This precipitation method will precipitate polymers that are longer than about 10 nucleotides. This general principle, the selective precipitation of a polymer, is widely used to assay enzymes that either degrade or polymerize polymeric molecules.

This assay will detect RNA polymerase in crude extracts despite the presence of the many enzyme inhibitors. The same assay may be used to follow enzyme activity during a purification procedure, but you should be aware that the day-to-day reproducibility of the assay is less than perfect and can vary as much as two-fold from one day to the next.

PROCEDURE

1. For crude extracts, grow cells to about 1×10^9 cells/ml, but not into stationary phase. Pellet the cells from 5 ml and resuspend them in 0.3 ml of breaking buffer (0.05 M KCl; 0.005 M $MgCl_2$; 0.01 M Tris–HCl, pH 7.9; 0.005 M 2-mercaptoethanol).
2. Sonicate the cells on ice and then pellet the debris at 5000 × *g* for 10 min.
3. Make the reaction cocktail and the assay mix.

Assay Mix:

10 μl cocktail
2.66 μl UTP, 5 μmol/ml, 2 μCi/μmol, and therefore 10μCi/ml
5 μl of 1.5 mg/ml calf thymus DNA dissolved in 10^{-2} M Tris-HCl, pH 7.5
dH_2O to make 100 μl

Reaction Cocktail:

160 μl 1 M Tris-HCl, pH 7.9
40 μl 1 M $MgCl_2$
1 μl 15 M 2-mercaptoethanol
40 μl of a solution with 25 mmol each of ATP, GTP, and CTP
150 μl dH_2O

4. Incubate the assay mix at 37 °C for 5 min and then add the enzyme. Use 25 to 50 μl of the crude extract described above in an assay mix of 100 μl. A concentrated RNA polymerase solution can be diluted in the cell-breaking buffer described in step 1.
5. After an incubation of 10–20 min, precipitate the entire reaction by adding 2.5 ml of 5% TCA (trichloroacetic acid), 0.01 M sodium pyrophosphate at 0 °C. Here, as in other precipitations in which labelled polymer is to be separated from labelled triphosphates, the inclusion of pyrophosphate reduces the nonspecific binding of triphosphates and hence lowers the background of the assay.
6. Collect the precipitate on Millipore filters, 0.45 μm, which have been stored more than 24 h in 5% TCA, 0.01 M sodium pyrophosphate, 1 M KCl. The KCl also appears to reduce nonspecific binding of triphosphates.
7. After precipitation, rinse each filter with about 4 ml of the TCA-sodium pyrophosphate solution at 0 °C and then dry and count. Normal levels for crude extracts are 400 cpm above a background of 75 cpm. As in all assays a zero time reaction sample should be tested for background.
8. For assays of crude extracts, make the assay mix 5×10^{-4} M in potassium phosphate to inhibit polynucleotide phosphorylase (Burgess, 1969).

ARABINOSE ISOMERASE

We include the assay for arabinose isomerase not only because we have had extensive experience with it, but also because many of the steps that are required for its efficient use are also required by assays for other enzymes. These diverse procedures include the use of sugar substrates; the safe, rapid, and accurate dispensing of concentrated sulfuric acid; and a rapid method for sealing the tops of hundreds of test tubes. The sealing prevents evaporation from the enzyme reactions during extended incubation. Finally, the heart of the assay, the cysteine–carbazole test, is fairly specific for a broad range of ketoses and for this reason has become the basis of many carbohydrate assays (Dische and Borenfreund, 1951).

Arabinose isomerase is the easiest arabinose operon enzyme to assay. The procedure can be performed in 45 min and it can be used to assay either pure enzyme, cell extracts, or whole permeabilized cells. In the assay the amount of L-arabinose converted to ribulose is proportional to the amount of added isomerase. The ribulose concentration is then measured with the cysteine–carbazole method.

The sensitivity of this assay is moderate, but with concentrated cells and extended incubation times the basal level of isomerase, of roughly 20 monomers/cell, will produce a response more than 20-fold above the background seen in an arabinose deletion strain. The reproducibility can be as good as 10% root mean square deviation.

STOCK SOLUTIONS

1. Double-strength assay mix: 20 ml 1 M Tris–HCl, pH 7.5; 15 ml 20% (wt/vol) L-arabinose; 60 ml H_2O; 5 ml 2 mg/ml chloramphenicol. The solution will last longer if it is kept sterile and stored in the refrigerator. Filter sterilize. Do not autoclave this or any other arabinose solution used in this assay because autoclaving will lead to an excessively high background.
2. 0.02 M $MnCl_2$: 0.25 g $MnCl_2$ or 0.396 g $MnCl_2$–4 H_2O per 100 ml H_2O.
3. 0.12% carbazole in 95% ethanol. Carbazole can be recrystallized from hot ethanol to reduce background.
4. 70% H_2SO_4: 450 ml concentrated H_2SO_4 added to 190 ml distilled H_2O.
5. 0.1 N HCl.
6. 1.5% cysteine–HCl. This should be made just prior to performing each set of assays because cysteine oxidizes to cystine during storage in solution.
7. Reaction mix: equal volumes of the double-strength assay mix and the $MnCl_2$ solution should be mixed no more than 2 h before use.
8. Toluene.

PROCEDURE

1. Cells can be used in any state of growth and in any medium. They can be used as they are or, if isomerase levels are low, they can be concentrated and then used. If cells are to be concentrated, spin them down, carefully

remove all the supernatant (a Pasteur pipette connected to an aspirator works well), and then resuspend them in a small volume, typically 0.5 ml, of 0 °C reaction mix. Make the cells permeable by adding 5 μl of toluene per 0.5 ml of cells and vortexing for 15 sec. The toluene does not need to be removed for subsequent steps.

2. The required volume of toluenized cells or enzyme sample is added to two test tubes each containing sufficient reaction mix so that the final volume is 100 $\pm$ 30 μl. Since the assay measures the amount of ribulose synthesized during incubation, each assay must have a blank sample that is not incubated. Hence, the second tube. The preceding steps are best done in an ice bath.
3. Add 0.1 N HCl to the blank sample so that its final volume is 1.0 ml. The HCl will prevent further enzyme activity. This tube can be left at room temperature while the other tube incubates at 37 °C.
4. Incubate the reaction tubes at 37 °C. The production of ribulose by isomerase is linear with incubation to about 24 h. If concentrated cells are being used, it may be best to shake them during the incubation to prevent clumping. Stop the reaction by adding HCl, as above.
5. After the reaction is halted, add 0.1 ml of freshly made 1.5% cysteine–HCl stock solution and mix.
6. Add 0.1 ml of 0.12% carbazole in 95% ethanol stock solution. The carbazole addition is most easily and accurately done with a mechanical micropipette (Eppendorf or Pipetman).
7. Immediately add 3 ml of 70% (by volume) H_2SO_4. It is best to add the sulfuric acid with a Lab Industries (Berkeley, Calif.) repipette. Although the usual repipette is adequate, special pipettes with weighted glass valves can be obtained from Lab Industries. These are more reliable for pipetting concentrated sulfuric acid. Don't worry about using the regular ones; if they don't work, you will know it. It is important that the H_2SO_4 be added to the assay tube immediately, that is, less than 3 sec after the carbazole addition; otherwise, excessive and variable amounts of a light-scattering precipitate will form. If the order of additions is reversed and H_2SO_4 precedes the carbazole the slight murkiness is avoided but the assay's reproducibility is lessened. Vortex briefly after adding the carbazole and very thoroughly after adding the acid. Usually 5–6 quick but interrupted vortexes suffice for thorough vortexing.
8. Wait 20 $\pm$ 0.5 min and measure the absorbance at 550 nm of both the incubated sample and the unincubated blank. The absorbance of the sample may be measured by transferring to a cuvette and measuring in a high precision spectrophotometer. However, the viscosity and general noxiousness of concentrated sulfuric acid plus the inherent variability in this assay make such a procedure a waste of time. A Bausch & Lomb Spectronic 20 spectrom-

eter is adequate for these measurements and has the added convenience that the measurements can be made right in the 13 × 100 mm reaction tubes. If these tubes are being used, rotate them in the light path to get a minimum absorbance. This will minimize the effect of tube imperfections. When doing many assays, use tubes from a single batch of 13 × 100 mm test tubes. Different batches of tubes frequently have slightly different diameters.

COMMENTS

1. The assay is linear with enzyme concentration when pure isomerase is used, but when toluenized cells are assayed it is rather nonlinear above a final OD_{550} of about 0.5.
2. The assay is very sensitive to changes in volume of the ribulose sample and the volume of H_2SO_4 added to that sample. A volume of 0.5 or 1.5 ml in the sample after the addition of HCl changes the final absorbance by 50–100%. A 10% deviation from the recommended addition of cysteine or carbazole will produce only a 2% deviation in the absorbance. The precision of the Lab Industries repipette has made any fluctuations in the volume of added H_2SO_4 irrelevant. Ribulose excreted into the medium by $AraB^-$ mutants growing in the presence of arabinose will interfere with the assay (Englesberg, 1966). This excretion can be used to score for such mutants and can also be used to synthesize ribulose.

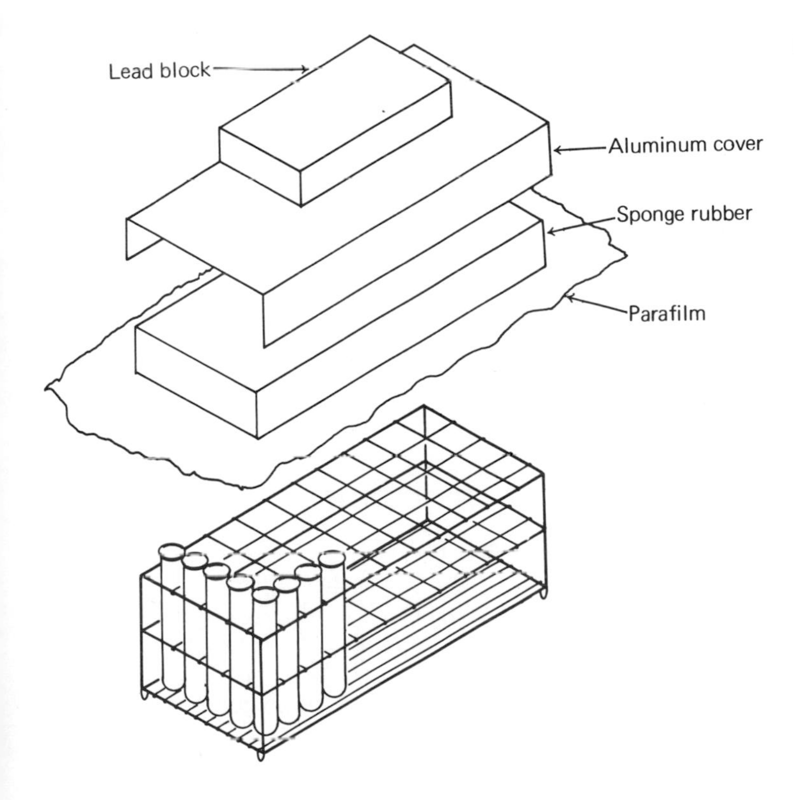

Figure 3.1. A method for sealing the tops of an entire rack of test tubes.

3. Tubes incubated for longer than 2 h should be tightly sealed to minimize evaporation. A test tube rack full of assay tubes may be rapidly sealed by covering the tops of the tubes with a sheet of Parafilm and holding it in place with a piece of sheet metal the size of the rack, to which 0.5-in. foam rubber has been glued (Fig. 3.1). Small lead weights placed on the sheet metal will ensure a good seal of the tubes.
4. In this cysteine–carbazole assay 10^{-7} mol of ribulose will produce an absorbance on the Spectronic 20 of 0.3, which is approximately equivalent to an $OD^{1\,cm}_{550\,nm}$ of 0.5 on a Zeiss PMQII. A fully induced or high level constitutive strain produces sufficient isomerase when grown to stationary phase in YT (about 2×10^9 cells/ml) so that 10 μl of toluenized cells incubated at 37 °C for 10 min will produce an OD of about 0.1. Convenient units are as follows:

$$\text{Units per cell} = \frac{2 \times 10^{12} \times OD_{550}\ \text{(Spectronic 20)}}{\text{minutes of incubation time} \times \text{number of cells in assay}}$$

A fully induced strain contains about 1000 units/cell when grown on minimal medium. The basal level is 0.5–10 units/cell. On the basis of data cited by Patrick and Lee (1968), this level of enzyme is 1.25×10^4 monomers or 2.1×10^3 molecules of the hexameric enzyme.

LYSOZYME

Historically the assay for phage lambda's endopeptidase, lysozyme, was important in measuring late gene activity of the phage (Dambly et al., 1968). It was later important in the study and purification of the lambda phage N and Q gene proteins (Greenblatt, 1972; Schechtman et al., 1980). This assay represents a class of assays which use an ill-defined substrate. As in all members of this class, the ill-defined substrate forces the assay's validity to rest on empirical findings rather than on theory.

The assay is a modification by Schleif et al. (1971) of that described by Dambly et al. (1968) and measures the rate of turbidity decrease in a suspension of sensitized whole cells.

PROCEDURE

1. Prepare the assay substrate from an overnight culture of the *E. coli* K-12 strain, C600 (Bachmann, 1972), in YT medium. Inoculate 100 ml of YT with 0.05 ml of the overnight culture and grow at 37 °C to an OD_{550} of 0.75 ± 0.25. This will take approximately 4.5 h. Pellet the cells and resuspend them in 50 ml of room temperature 0.10 M EDTA, neutralized with NaOH to pH 8.0. Leave at room

temperature for 5 min. Pellet the cells at 4 °C, resuspend them in 40 ml of 0.01 M potassium phosphate, pH 6.8, and store on ice. These cells will remain sensitive to lysozyme for at least 4 h. The OD_{600} of the resultant cells should be about 1.2. Adjust the substrate concentration to an OD_{600} of between 0.4 and 0.8 by diluting with the same phosphate buffer.

2. Prepare an enzyme sample by growing a strain that is lysogenic for the heat-inducible lambda, CI_{857}, in YT at 35 °C to an OD_{550} of 0.5. Induce by shifting the temperature to 42 °C for 15 to 20 min and then grow at 35 °C for 45 min after beginning induction. Chill 2 ml of this culture on ice and then sonicate it on ice. Spin out the debris and use the supernatant as the enzyme sample for the assay.
3. Prepare the assay reaction mixture by first warming several ml of substrate to room temperature. Add 10 μl of the enzyme sample to 1 ml of the sensitized cells in a cuvette and mix completely by shaking with Parafilm over the cuvette top. At 1-min intervals read the OD_{600} of this reaction mix and of a control cuvette that contains all the ingredients except the enzyme. Both readings should be compared to the OD_{600} of a cuvette with H_2O only, or of no cuvette at all. A volume of 10 μl of supernatant from CI_{857} changes the OD_{600} from the initial value of 0.5 to about 0.1 after an 8-min incubation, while the "cells only" control sample changes from 0.5 to about 0.4 after 8 min.

COMMENTS

1. Cells spontaneously lyse at a rate proportional to the number present; that is:

$$\frac{dOD}{dt} = -k_1 \cdot OD.$$

They also lyse (a lysed cell has no OD_{600}) at a rate proportional to the amount of lysozyme present, E. Thus, the total rate of lysis can be described by

$$\frac{dOD}{dt} = -k_1 \cdot OD - E \cdot OD$$

and the enzyme concentration is thus given as

$$E = \frac{-\Delta \ln OD}{t} - k_1.$$

When OD is plotted on semilog paper, the lysozyme present is proportional to the increase in the rate of cell lysis due to the addition of supernatant, that is, it is proportional to the distance indicated as Δ in Figure 3.2. Note that about 10% of the cells are not sensitive to lysozyme, so the OD levels off at about 0.05. For this reason, quantitation of lysozyme should be done within the linear portion of the plot.

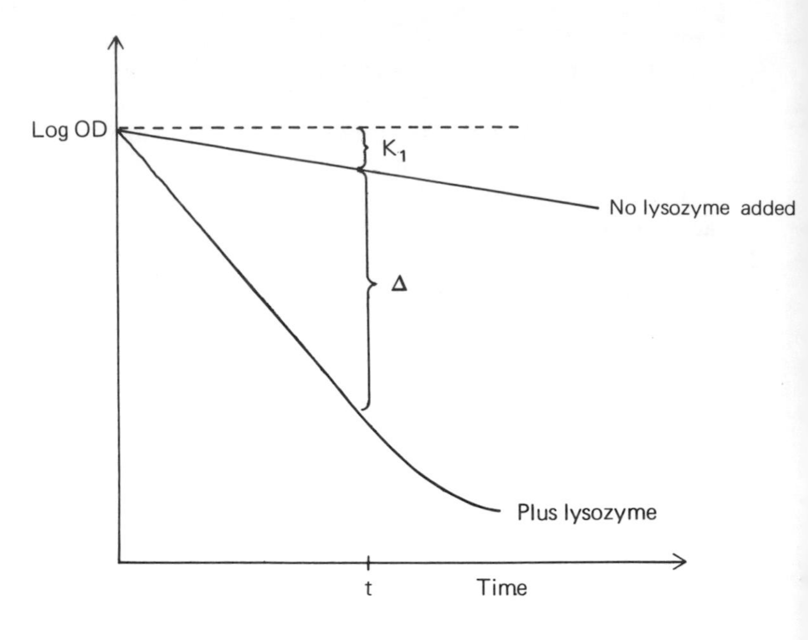

Figure 3.2. Kinetics of loss of cell absorbance in the lambda endopeptidase (lysozyme) assay.

2. The assay is linear with lysozyme concentration for at least a 50-fold dilution from the above enzyme sample. Useful signals can still be detected upon further dilution. At least 100 μl of supernatant enzyme sample can be added to the sensitized cells without noticeably affecting the response. At high concentrations of lysozyme the slope is not proportional to the amount of enzyme added. Therefore check the linearity by dilution.
3. Chicken egg-white lysozyme can also be used to normalize this procedure and to check the suitability of sensitized cells. An OD_{600} decrease from 0.5 to 0.15 in 8 min is typical for 0.5 μg of pure egg-white lysozyme if the substrate cells are adequately sensitive.

RIBULOKINASE

Many enzymes transfer a phosphate to a small molecule. Here we describe our experience with one of these, ribulokinase. We present this ribulose ribulokinase assay in detail because it is a more general approach than the usual sugar kinase assays and it provides a less elaborate technique for separating the reaction components. We expect that simple modifications will be sufficient for its use with many similar enzymatic reactions. As an example, an enzyme of broader interest, galactokinase, probably can be assayed by essentially the same procedure. Although we have not had extensive experience with galactokinase, the literature indicates (Parks et al., 1971) that the assay we describe could be used.

In order to assay kinases it is necessary either to separate the unphosphorylated and phosphorylated small molecules or to separate the phosphate and the phosphorylated small molecules. The ribulokinase assay described below separates the former pair by virtue of the binding of phosphate and phosphorylated small molecules to DEAE paper. Such a method is convenient since it usually permits the use of long half-life labels such as ^{14}C rather than the rapidly decaying radioactive phosphorus in phosphate.

Two basic methods exist for using DEAE paper to separate phosphorylated small molecules. In the first the reaction mix is filtered through DEAE paper, and in the second the mix is chromatographed on DEAE paper. For the ribulokinase assay two problems were encountered when using the filtration method (Schleif et al., 1973). First, ribulose phosphate did not quantitatively bind to the paper and washing the paper to remove unphosphorylated substrate also eluted some of the phosphorylated product. Second, no reasonable amount of washing could reduce the high background due to binding of the unreacted substrate. These problems were not encountered when a chromatographic separation was used. Developing the DEAE paper chromatogram with water moves all of the unreacted substrate well away from the origin, completely separating it from the phosphorylated product which hardly moves at all.

This technique gives very low backgrounds and hence high assay sensitivity. However, it does generate a new problem. The substrate must be free of radiolabelled impurities which would remain at the origin during chromatography. In the ribulokinase assay the requisite purity was obtained by purifying the substrate through a DEAE column.

The L-ribulokinase assay is very sensitive and can measure the basal levels of ribulokinase in whole permeable cells, cell extracts, or purified protein. In the assay [^{14}C]ribulose is converted to [^{14}C]ribulose-5-phosphate, and the sugar phosphate is then isolated and counted. In practice it is much cheaper to add [^{14}C]arabinose and arabinose isomerase to the assay than to obtain [^{14}C]ribulose (Schleif et al., 1973). The isomerase in the reaction mix converts the [^{14}C]arabinose to [^{14}C]ribulose, the substrate of ribulokinase.

In addition to describing a general kinase assay the procedure also presents a convenient general method for the spotting and developing of a large number of reaction mixes. Fifty may be done in an afternoon.

PROCEDURE

1. If whole cells are to be assayed for kinase they can be grown in any medium and used in any state of growth. However, if arabinose has been used to induce a culture, then it *must* be removed before assaying. In general, spin down 3.5 ml of cells and resuspend in an equal volume of M10 medium, spin down again, and then resuspend either in 0.01 M Tris-HCl, pH 7.6, 0.1 mM DTT, if enzyme levels are high, or in 100 μl assay mix (see Assay Mixture, below), if levels are low. Add 5 μl of toluene, vortex for

20 sec, and allow to sit on ice for one half-hour. In contrast to the brief period of toluenization required for the β-galactosidase or arabinose isomerase assays, the ribulokinase assay requires this longer interval of toluenization.

2. Dilute the toluenized cells or enzyme solution through enzyme assay mix and adjust reaction volumes to 100 μl.
3. Incubate the tubes at 30 °C for the required length of time, 30 min to 24 h.
4. Insert a folded strip of DEAE paper into a reaction tube (see below). It will imbibe all the reaction mix and can then be hung on the chromatography rack. It is best to begin the chromatography before the spot has dried; otherwise a higher background will result. Develop the strip upward with water until the front is 1 cm from the top. Remove the strips from the water, dry, and count as explained below.
5. A more standard, but more involved, procedure for precipitating other sugar phosphates is to precipitate ribulose phosphate with barium. Add to the assay reaction 5 μl 200 mM glucose-1-phosphate (as carrier), 100 μl 1 M $BaCl_2$, and 1 ml absolute ethanol. Store on ice for 20 min. Filter the reaction through a glass fiber filter (GF/C, Whatman) and then wash the filter with cold 80% ethanol. Count the filter in water-compatible scintillation fluid.

ASSAY MIXTURE

One ml of assay mix contains the following:

Volume	Component	Final Concentration
0.56 ml	H_2O	
0.1 ml	0.5 M K^+-phosphate, pH 7.8	0.05 M
10 μl	0.1 M K^+-EDTA	0.001 M
30 μl	1 M Mg^{2+}-acetate	0.03 M
50 μl	160 mM ATP	0.008 M
5 μl	1 M NaN_3	0.005 M
20 μl	0.1 M DTT	0.002 M
0.17 ml	2 mg/ml chloramphenicol	0.35 mg/ml
50 μl	10 mg/ml streptomycin	0.5 mg/ml
10 μl	1 M NaF	0.01 M
20 μl	isomerase, 2×10^{12} molecules/μl, free of ribulokinase	
10 μl	[^{14}C]-L-arabinose, 10,000 cpm/μl, 10 mCi/mmol	
0.4 μl	20% L-arabinose	0.008%

COMMENTS

1. Purifying [^{14}C]arabinose on DEAE reduces the background. Dry the [^{14}C]arabinose in a tube, dissolve it in

water, and then pass it over a 1-ml column of DE52 (Whatman) which has been equilibrated with 0.1 M Tris-HCl and flushed with water.

2. The assay is linear for 18–20 h at 30 °C.
3. For cells with a basal level of ribulokinase, 10^9 cells incubated at 30 °C for 1 h yield 800–900 cpm. Fully induced cells yield more.
4. It is useful to keep a stock of an extract of a strain without ribulokinase activity, *ara* $C^cB^-A^+$, to use as a control when running the assay. Extracts stored in 50% glycerol in the freezer seem not to lose activity for at least several months.
5. In general, a maximum of 30–50% of the total counts added to the reaction can be converted to ribulose phosphate.
6. The most convenient method (Figure 3.3) for spotting the paper is to fold the strip in half so that the folded strip is 0.75 × 17 cm. Then fold this over 3 cm from one end. This folded end is inserted into the tube of reaction mix and quickly absorbs all the liquid. Remove the paper from the tube and unfold it before performing the ascending chromatography.

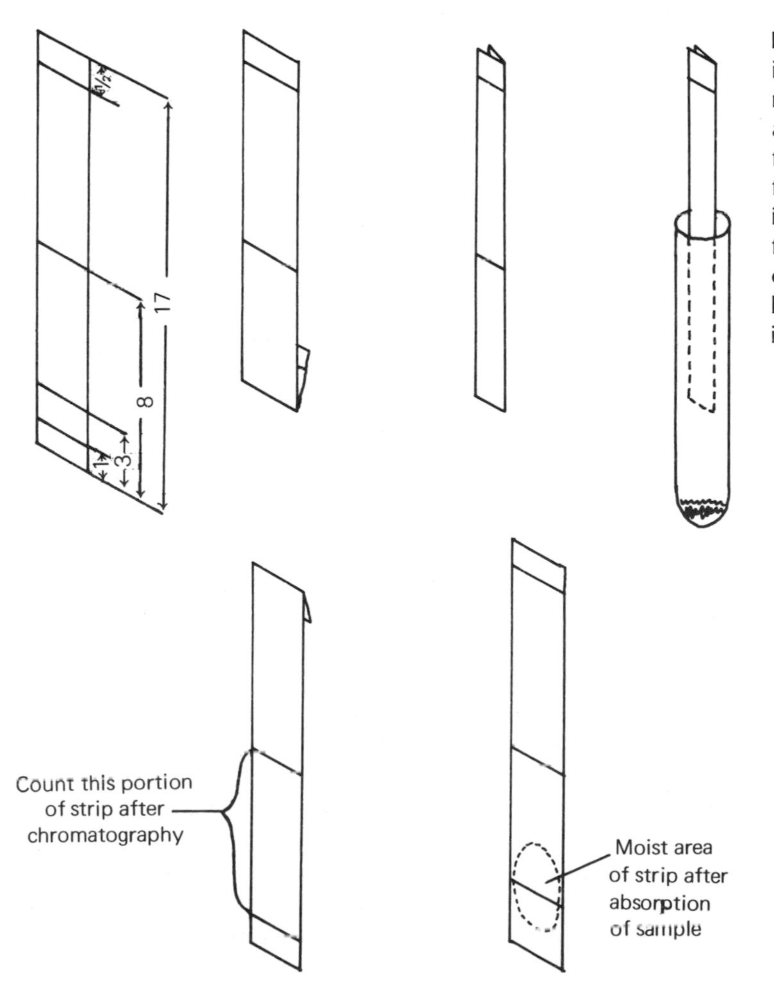

Figure 3.3. The dimensions and folding of the DEAE filter paper for the ribulokinase assay. The dimensions are shown in cm; the first fold 3 cm from the bottom: the second fold folding the paper in half: and then inserting the paper in the tube and folding. Finally, the paper is folded 1 cm from the top to permit it to be hung from the apparatus for ascending chromatography.

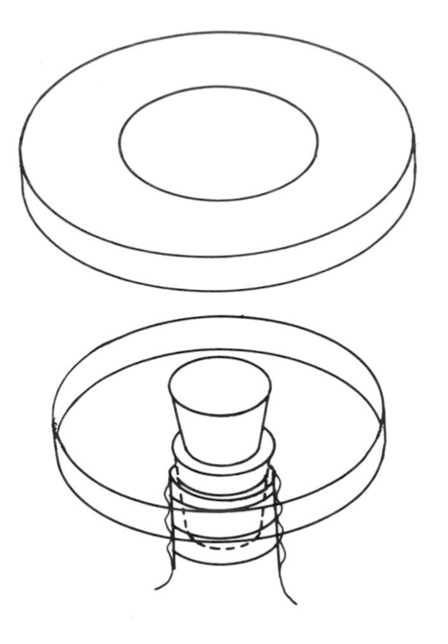

Figure 3.4. The ascending chromatography apparatus.

7. A convenient way to do the ascending chromatography is the following. Insert a stopper into the top of a milk dilution bottle and glue the top of the stopper to the bottom side of a petri plate bottom. Glue a second stopper to the top side of the petri plate bottom (Figure 3.4). The second stopper serves as a handle. Hang the DEAE strips over the lip of the petri plate bottom and then place a petri plate top over the bottom so that it will hold the strips in place. A hole must be punched in the top to permit it to fit without interference from the handle (Figure 3.5). The hole can be made with a piece of heated metal of the right size. Up to 10 strips can be chromatographed on one bottle. Chromatograph the strips by placing the bottle and strips in the lid of a larger petri plate and then add water until it just touches all the strips. Put a 3-liter beaker over the bottle with strips to enclose the system so that the water does not evaporate from the strips as rapidly as it rises (Figure 3.6). Allow the water to rise until it just reaches the lip of the top petri plate. This will take 60 to 120 min. Dry the strips, cut off the bottom 9 cm, and count in 2.5 ml of scintillation fluid.

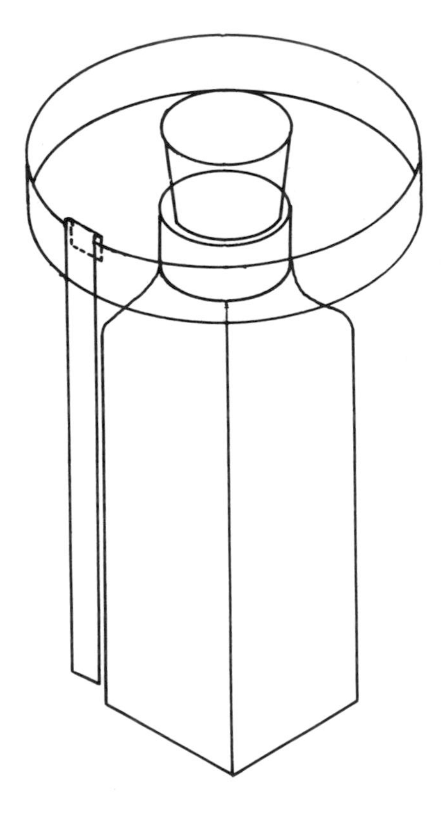

Figure 3.5. A chromatography strip is hung from the bottom petri plate of the chromatography apparatus.

E. COLI-COUPLED TRANSCRIPTION-TRANSLATION SYSTEM

Cellular molecules that influence the specific transcription and translation of a particular gene can be identified and assayed by one or another of several techniques that share a common strategy. The strategy uses a cellular extract that has the basic components for transcription and/or translation, but lacks both the specific factors to be assayed and the end product to be measured. The gene or mRNA to be transcribed or translated is added to this extract and the suspected specific factors are also added. After incubation, the amount of end product synthesized is measured to determine the activity of the suspected factors (Greenblatt and Schleif, 1971). A contemporary illustration of this strategy would be an examination of the regulation of an *E. coli* ribosomal protein operon. The cloned operon DNA (p. 129) could be added to the Zubay coupled transcription-translation system (see below) and ribosomal proteins could be added to determine if they exert the proposed (Yates et al., 1980) feedback inhibitory effect on the synthesis of the proteins. Measuring this effect would be greatly simplified by fusing the galactosidase gene to the operator-promoter region of the cloned ribosomal protein operon because this gene product is easily and sensitively assayed (p. 43).

There are several in vitro transcription and translation systems that can be used in this way. Two eukaryotic translation systems are described below (p. 161). In this section the prokaryotic coupled transcription-translation system mentioned above is described. This is a well-characterized system which has become increasingly powerful as gene isolation and manipulation techniques have improved. Several research groups have contributed to the development of this *E. coli* system, but the methods described by Zubay et al. (1970) have proved to be most useful.

Four components are necessary for this system. The first is known as S-30. It is a crude cell extract which provides the enzymes, ribosomes, and most of the other cellular material necessary for transcription and translation. As suggested above, the strain from which the S-30 is made must lack both the gene products to be assayed and those which will be synthesized. The second major component in the assay is the DNA template. This is usually either a plasmid DNA or a lambda phage DNA that contains the gene of interest. The third component is a mixture of salts, buffers, precursors, and so on that will be called the super reaction mix (sRM). The fourth component, called the "other extract," is a cell extract which is the source of the protein or factor to be assayed. In the example mentioned above the "other extract" would be purified ribosomal proteins.

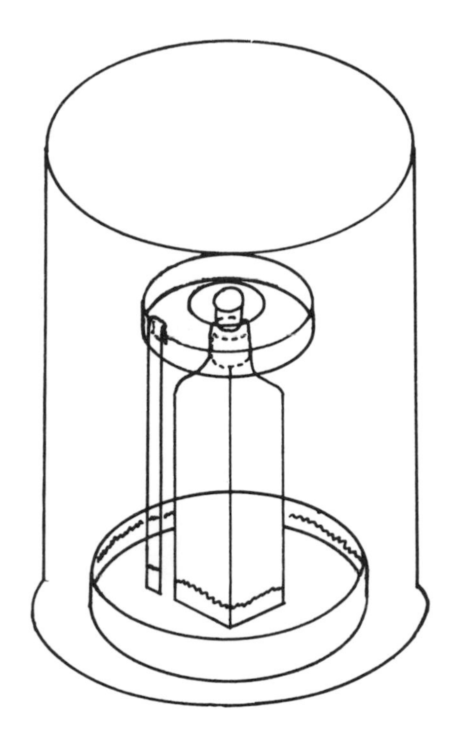

Figure 3.6. The assembled chromatography apparatus with the hanging strips resting in a small dish of water. The entire apparatus is enclosed in a 3-liter beaker.

PROCEDURE

1. Growing Cells

Add 0.6 ml of an overnight YT culture of a suitable strain to each of nine 2-liter flasks containing 600 ml of nutrient broth, 15 ml of 20% glucose, 6 ml of vitamin B_1 (1 mg/ml), and 600 ml deionized, distilled water. Glucose and vitamin B_1 solutions should be autoclaved separately and added to the nutrient broth flasks. Cells are grown at 33–34 °C with shaking at 320 rpm to a final OD_{550} of 1.0 ± 0.15. The cell-doubling time is approximately 35 min and total growth time is roughly 3–4 h. The OD of each flask should be monitored. When the desired OD_{550} is reached, swirl the flask in an ice water bath to chill the cultures rapidly. Perform all subsequent steps in the cold with prechilled buffers and rotors as *rapidly* as possible. Centrifuge cells at 5500 rpm (7500 × *g*) for 10 min in the GSA rotor of a Sorvall centrifuge. Three centrifuges are required. Use large clear plastic bottles. Resuspend the combined cell pellets twice in 21 ml of buffer W (see Recipes, below) and centrifuge at 16,000 rpm for 10 min in the Sorvall SS-34 rotor (20,000 × *g*). After the first 16,000-rpm centrifugation, disperse the pellet first with a spatula to assist resuspension, and then vortex.

2. Grinding Cells

Place approximately 4 g of washed cells in a cold (0–4 °C) mortar and cover with twice the cell weight of cold alumina (p. 157). Grind cells in the cold room for 3.5 min. Wear plastic gloves. A popping sound should be heard after 1–2 min of grinding and the final suspension should be a sticky white paste. Suspend the paste in buffer B, 0–4 °C. Add twice as many milliliters of buffer B as grams of cells you first added to the mortar. Add buffer B in thirds, mixing each third thoroughly with the paste before the next addition. With the aid of a spatula and a finger, pour the suspension into a cold Sorvall tube. Also prepare a balance tube containing twice the cell weight of alumina and water. Centrifuge the suspension at 16,000 rpm (20,000 × g) for 20 min in the SS-34 rotor at 0–4 °C. Pour all of the supernatant off and very gently mix to homogeneity. The supernatant fluid is light-sensitive and should be kept in the dark as much as possible.

3. Preincubating

Add the supernatant to the appropriate volume of preincubation mix. Gently mix the tube contents by covering with Parafilm and inverting 3 times. Preincubate the mixture for 80 min at 37 °C in the dark without shaking. Transfer the preincubated S-30 to a #20 dialysis sack and dialyse at 0–4 °C against buffer F for a total of 4 h in the dark, with one change of buffer after 2 h. Each change of buffer involves 200 ml. The dialysis is most conveniently done in an Erlenmeyer flask placed in a covered ice bucket on a magnetic stirring motor.

Aliquot the dialysed S-30 into new disposable glass test tubes (10 × 75 mm) in 200-μl portions and quick-freeze in a dry ice–acetone bath. Store them at −70 °C until use. A biuret protein determination on the S-30 should yield approximately 15 mg/ml.

4. Other Extract

Once this extract is made it is best to dialyse it against the "other extract buffer" before it is added to the reaction mix.

5. The Synthesis Reaction

Below are given the components for a Zubay synthesis to be performed in a total final volume of 1.0 ml. The volume of the super reaction mix (sRM), the S-30, and the other extract to be added will vary depending on the assay. When making the sRM, the order of addition of components should be as listed.

The standard addition volumes for the first four components of the sRM depend on the volumes of the S-30 and the other extract added to the 50-μl synthesis. If the added volumes of the S-30 and the other extract will vary within a given experiment, a standard volume should be decided upon for the S-30 plus S-30 buffer and for the other extract plus other

Standard volume for sRM	Component	Final concentration in synthesis reaction
variable	2 M Tris-acetate, pH 8.2	44 mM
variable	5 M K^+-acetate	55 mM
variable	0.1 M DTT	1.4 mM
variable	1 M Mg^{2+}-acetate	variable
13.5 μl	2 M NH_4^+-acetate	27 mM
8.0 mg	phosphoenolpyruvate	8 mg/ml
4.4 μl	all amino acids, each at 50 μM	0.22 mM
10.0 μl	vitamin stock	—
5.5 μl	100 mM UTP (neutralized with KOH)	0.55 mM
5.5 μl	100 mM CTP (neutralized with KOH)	0.55 mM
5.5 μl	100 mM GTP (neutralized with KOH)	0.55 mM
22.0 μl	100 mM ATP (neutralized with KOH)	2.2 mM
20.0 μl	tetrahydrofolic acid (calcium leucovorin, injection, Lederle Laboratories)	—
150.0 μl	500 μg/ml template	75 μg/ml

extract buffer. For example, the added volumes of S-30 and S-30 buffer might always total 15 μl and the volumes of the other extract and other extract buffer might always total 10 μl. Adjust the volumes of the variable components of the sRM to give the desired final concentrations of all components. Use double-distilled water to bring the final Zubay synthesis volume to 50 μl. Set up the reaction in new 10 × 75 mm disposable glass test tubes. The order of addition of components is water, buffers, sRM, S-30, and other protein extract. The sRM and the S-30 may be premixed. Gently mix the tubes after the addition of each component, and incubate the synthesis mixture for 30 min at 30 °C with moderate shaking in the dark. When using an S-30 for the first time, the magnesium optimum for the synthesis should be determined by varying the Mg^{2+}-acetate. The optimum is usually found between 14 and 20 mM. Usually an amount of S-30 to yield 7.5 mg/ml in the final reaction volume is optimum. Following synthesis, the reaction mix can be assayed for the synthesized gene product by enzyme assay, gel electrophoresis, or biological assay.

RECIPES

All buffers should be made in double-distilled water (see Appendix I, Commonly Used Recipes). The Tris-acetate, Mg^{2+}-acetate, KCl, and K^+-acetate solutions should be filtered through boiled (see Appendix I) 0.45-μm Millipore filters before use. It is very important that all solutions which come into contact with S-30 be made and stored in glass that has been washed with chromic acid-sulfuric acid, rinsed with 0.01 M Na^+-citrate, and then extensively rinsed with deionized, distilled water. This includes the containers for all stock solutions, the centrifuge bottles, and the test tubes.

	Volume	Component	Final Concentration
Buffer W	0.25 ml	2 M Tris–acetate, pH 7.8	10 mM
	0.7 ml	1 M Mg^{2+}-acetate	14 mM
	1.5 ml	2 M KCl	60 mM
	0.05 ml	0.1 M dithiothreitol	0.1 mM
	dilute to 50 ml with H_2O		
Buffer B	50 μl	2 M Tris–acetate, pH 7.8	10 mM
	140 μl	1 M Mg^{2+}-acetate	14 mM
	120 μl	5 M K^+-acetate	60 mM
	10 μl	0.1 M dithiothreitol	0.1 mM
	9.68 ml	H_2O	
Preincubation mix	0.35 ml	2 M Tris–acetate, pH 7.8	
	19.6 μl	1 M Mg^{2+}-acetate	
	56 μl	100 mM ATP	
	0.84 ml	buffer B containing 19.6 mg Na_3^+-phosphoenol-pyruvate	
	0.07 ml	0.1 M dithiothreitol	
	0.14 ml	all amino acids each at 500 μM	
	7 μl	phosphoenolpyruvate kinase, 10 mg/ml	
	0.714 ml	H_2O	
	add 0.314 ml/ml of supernatant obtained		
Buffer F	2.5 ml	2 M Tris–acetate, pH 7.8	10 mM
	5.6 ml	1 M Mg^{2+}-acetate	14 mM
	4.8 ml	5 M K^+-acetate	60 mM
	4.0 ml	0.1 M dithiothreitol	1 mM
	383.0 ml	H_2O	
Vitamin stock	2.7 mg/ml	FAD	
	2.7 mg/ml	TPN	
	2.7 mg/ml	pyridoxine	
	1.1 mg/ml	para-aminobenzoic acid	
Other extract buffer	0.5 ml	2 M Tris–acetate, pH 7.8	10 mM
	1.2 ml	5 M K^+-acetate	60 mM
	1.4 ml	1 M Mg^{2+}-acetate	14 mM
	0.5 ml	0.2 M K^+-EDTA, pH 7.0	1 mM
	1.0 ml	0.1 M DTT	1 mM
	94.9 ml	H_2O (or dilute to 200 ml)	

Chapter 4

Working with Proteins

Most research in molecular biology uses proteins at every turn. Frequently it is necessary to purify or partially purify them, and in almost all experiments it is necessary to use an enzyme for one purpose or another. Proteins are remarkably labile structures, probably resulting from the fact that the equation of forces between those tending to destroy structure and those serving to maintain a protein's structure is only slightly tipped toward maintenance of structure (Schultz and Schirmer, 1978). Proteins are sensitive to pH, monovalent and divalent ions, specific ions and specific metals, temperature, and protein concentration. Thus it is important to pay close attention to buffers when working with proteins. What appears to be an awkward step in a certain procedure is likely to be included because it is the only known method that works. In most cases proteins are more stable at low temperatures than at room temperature. Thus most purification steps are performed at 0–4 °C and enzymes are stored at temperatures as low as possible.

The remarkable diversity in properties of proteins precludes writing any short set of generally applicable instructions. The material which follows contains some of the specific recipes which are often used in conjunction with protein purifications. This list is in no way exhaustive and the best strategy for a beginner in manipulating proteins would be to read the procedures which have been used for the purification and manipulation of a number of different proteins. The series, *Methods in Enzymology,* by Academic Press contains much information along these lines.

Some of the procedures used in protein purifications or in the manipulation of proteins are used at several stages. However, a few of the procedures often follow one another in a standard order. This chapter is organized with this tendency in mind. Once cells have been opened (p. 15), many times it is useful to perform a centrifugation or a precipitation step to remove ribosomes and other large cellular components. The details required for this vary, depending on the viscosity of the resulting cell lysate and centrifugation facilities available. Typically, a centrifugation at 30,000 rpm in the Beckman angle 30 rotor, $100{,}000 \times g$, for 90 min suffices. Removal of ribosomes can be monitored by absorbance at 260 nm. Often the next step is removal of the remaining nucleic acids by phase partition (p. 64). Then an ammonium sulfate fractionation (p. 62) is usually performed. This is followed by column chromatography steps (p. 68). Occasionally, ammonium sulfate precipitation can be used to perform a crude fractionation of proteins at the same time that it removes nucleic acids from the desired fraction.

In the middle of the chapter we describe techniques that are used at many stages of a purification project. These techniques include determining protein concentrations and concentrating protein solutions. At the end of the chapter we describe several gel electrophoresis techniques, valuable tools for evaluating steps of a protein purification.

AMMONIUM SULFATE PRECIPITATION OF PROTEINS

Crude extracts can be roughly fractionated by ammonium sulfate precipitations. Typically, about a 5-fold increase in protein purity can be obtained and often extraneous DNA or tRNA is removed as well. The basic principle is that different proteins are precipitated by different concentrations of salts. Most often ammonium sulfate is used for precipitation. Our experience is limited to this precipitant. Presumably other highly soluble innocuous salts such as potassium phosphate could also be used. A few warnings are in order before using this technique. Be aware that a number of metal ions may contaminate ammonium sulfate and inactivate the protein. Therefore it pays to use enzyme grade ammonium sulfate. For large scale purifications crude ammonium sulfate can sometimes be used in the presence of EDTA to chelate these ions. Enzyme grade ammonium sulfate also requires much less care to maintain constant pH during precipitation than does crude ammonium sulfate. Note that proteins are often coprecipitated with other proteins and thus your enzyme may precipitate at different ammonium sulfate concentrations if preceding steps are changed. Similarly, the precipitation is a combined effect of the salt and the protein so that if the protein is appreciably diluted, higher concentrations of ammonium sulfate may be required for precipitation.

Few proteins from *E. coli* are precipitated at ammonium concentrations below 24% of saturation, most proteins precipitate at about 35%, and ribosomes precipitate at about 45%. Almost all proteins which can be precipitated have done so by about 55% of saturation. Most buffers at 0–4 °C are saturated by about 70.5 g/100 ml of ammonium sulfate. Finally, for purposes of calculating how to perform a series of steps of salt precipitation, it is convenient and sufficiently accurate to assume that the specific volume of ammonium sulfate is 0.5 ml. That is, if a gram of ammonium sulfate dissolves in a buffer, the volume of the buffer is increased by 0.5 ml.

PROCEDURE

Given below is a typical procedure for isolating a 25–33% "ammonium sulfate cut."

1. Measure the volume of the extract. For purposes of the example, assume it is 250 ml.
2. To bring the extract to 25% of saturation in ammonium sulfate, $250 \times 0.25 \times (70.5/100) = 44$ g must be added. To a beaker of extract stirring in ice water, slowly add the ammonium sulfate over an interval of 5–10 min. At several times during the procedure check that the pH of the buffer has not altered appreciably. Usually a drop or two of 1 M NaOH or KOH must be added per 10 g ammonium sulfate in order to hold the pH constant.
3. Stir for a set interval of time. Usually 20 min is adequate, but be aware that proteins will continue to precipitate with time. A protein "cut" with precipitated proteins removed will usually become cloudy with further precipitated proteins during 12 h at 4 °C.
4. Centrifuge down the precipitated proteins. Most often $10{,}000 \times g$ for 20 min is adequate.
5. The protein in the pellet can be resuspended and assayed for the desired protein. Such resuspended protein contains much ammonium sulfate. In the example we are considering, this pellet would be discarded.
6. Use the supernatant solution for the next step of ammonium sulfate fractionation. In our example, this solution is to be brought to 33% of saturation in ammonium sulfate. If the original 250 ml had been brought immediately to 33% of saturation, $250 \times 0.33 \times (70.5/100) = 58$ g of ammonium sulfate would have been added. However, 44 g had already been added to bring the solution to 25% of saturation. Thus an additional $(58 - 44) = 14$ g must be added. Usually the recovery of the supernatants is less than complete. Let us assume that Y ml rather than 272 $(250 + [44 \times 0.5])$ ml had been recovered from the first precipitation step. Then $(Y/272) \times 14$ g of ammonium sulfate are required to bring the supernatant to 33% of saturation. The second cycle of precipitation should be performed just like the first.

COMMENTS

A variety of schemes exist for determining the amounts of ammonium sulfate to add to produce the cuts desired.

When following someone else's procedure, follow the exact procedure given and do not modify it, for example, by calculating the volume of a saturated ammonium sulfate solution to use to replace a step where addition of solid ammonium sulfate is indicated. The actual degree of saturation of ammonium sulfate solutions depends on the buffer and temperature. The specific volume of ammonium sulfate also depends on the degree of saturation of the buffer. The approximate numbers we have given above are fine for reproducing work within a laboratory, but may differ importantly from the most refined numbers. Such differences are irrelevant, since in the end one should report *x* g ammonium sulfate to *y* ml of extract, and someone else will follow this procedure rather than ever bothering to calculate the degree of saturation produced. A typical set of ammonium sulfate cuts which might be tried to purify a new enzyme are 0–28%, 28–34%, 34–40%, and 40–60%.

REMOVING NUCLEIC ACIDS BY PHASE PARTITION

Usually, the first step in a protein purification procedure removes nucleic acids. Traditionally this is accomplished by precipitating the nucleic acids with cations such as Mn^{2+}, streptomycin, or protamine. However, these methods are unsatisfactory if subsequent steps will be assayed by in vitro protein synthesis, if the desired protein does not survive the precipitation conditions, or if the extract containing the protein is particularly high in nucleic acid. The latter may well be the case if multiple copies of lambda phage or a plasmid are being used to boost synthesis of the protein. In all of these cases a phase partition system is a good method for removing nucleic acids.

Dilute aqueous solutions of dextran 500 and polyethylene glycol 6000 form two-phase systems over a wide range of polymer concentrations. (See Albertsson [1960] for a phase diagram describing the distribution of the polymers between the two phases as a function of their relative concentrations.) For molecular biologists the utility of such systems lies in the fact that macromolecules such as proteins and nucleic acids will partition between the two phases in a manner critically dependent on their structure and the ionic composition of the phase system. Thus, liquid polymer phase partitioning provides a powerful, yet very gentle, means of fractionating mixtures of nucleic acids, proteins, or of nucleic acids and proteins. Phase systems can be used to separate DNA on the basis of strandedness, or to separate DNA from RNA (Alberts, 1967), but the descriptions here will be confined to their use in separating nucleic acids from proteins, and in purifying RNA- or DNA-bound proteins.

REAGENTS AND STOCK SOLUTIONS

Store the dextran solutions at 4 °C.

1. Dextran T500 from Pharmacia. There are reports of variation from lot to lot in this material, so it is wise to do a small scale check of a new lot.
2. Polyethylene glycol (PEG) 6000 (Carbowax 6000 [Union Carbide]). This is an industrial chemical and is easier to obtain in 100-lb quantities than in small packages. It is inexpensive.
3. Recipe A (D 6.4; PEG 25.6; 4 M NaCl): Heat about 500 ml of 4 M NaCl in distilled water to near boiling on a stirring hot plate. Slowly add 64 g of Dextran T500. After this is dissolved, add 256 g PEG 6000. Continue to heat and stir until the initially gummy mess dissolves and forms a homogeneous mixture. Do not forget that this is a two-phase system and will be very cloudy. Complete to a final weight of 900 g by adding 4 M NaCl. Alternatively, one can use 4 M NaCl buffered with a buffer of choice (0.05 M or less), and then complete the final weight to 1000 g.
4. Recipe B (D 6.4; PEG 25.6; no salt): Prepare this just as A, but use water instead of 4 M NaCl.
5. Recipe C (PEG 6.9%): Heat buffer containing 1, 2, or 4 M NaCl (see below) to 40–50 °C with stirring. Use warm buffer to make a 6.9% (wt/wt) PEG 6000 solution. Stir until all polymer dissolves.

Bulk Separation of Proteins from Nucleic Acids

This procedure can be used to remove DNA, high molecular weight RNA, and most tRNA from cell extracts. Oligonucleotides are not removed efficiently, so nuclease treatments are to be avoided except to reduce unmanageably high viscosity in large preparations. In this case DNase treatment at 2 μg/ml for 30 min at 0 °C reduces viscosity but still permits separation of nucleic acids and protein.

PROCEDURE

1. Prepare a crude extract by breaking cells with any of the usual methods. Spin out the debris at low speed, 5000 × *g*. The extract will not be completely clear. Do not use DNase treatment if at all possible. The extract should be kept at 4 °C.
2. While the extract is being made, put the concentrated stock solution of polymer in a warm water bath. Heat until the polymers become sufficiently liquid that they can be thoroughly mixed. Shake vigorously until all of the gummy bottom layer, the dextran phase, is mixed in. This is crucial.
3. Preweigh a beaker and stirring bar on a pan balance. Add

the extract and calculate its weight, W. Return the extract to the cold room or to an ice bucket. Then weigh out enough NaCl to make the extract 4 M NaCl. Since this is the same as 19.6% NaCl (wt/wt), the amount of NaCl to be added, G_{NaCl}, may be calculated as follows:

$$G_{NaCl} = \frac{0.196\ W}{1 - 0.196} = 0.243\ W$$

Add the salt to the extract while stirring to dissolve.

4. If the polymer had been made to 1000 g then add one-third weight of polymer mix per weight of salted extract to achieve the final desired concentrations of dextran and PEG. If the polymer mix was made up to 900 g, the required amount is 0.9/3 weights of polymer mix per weight of extract. Thus,

$$G_{polymer} = \frac{0.9}{3}(G_{NaCl} + W) = 0.3\,(G_{NaCl} + W) = 0.37\ W$$

Liquid of weight (0.1/3) ($G_{NaCl} + W$) = 0.041 W is then added to complete the system. The liquid could be water but preferably would be a 10-fold concentrate of the cell-breaking buffer.

5. Stir this completed mixture for 30 to 60 min at 4 °C.
6. Centrifuge at low speed, 5000 × g, for 10–15 min. After centrifugation, the tubes should contain a clear yellow upper phase above a cloudy brownish or grayish bottom phase. The bottom phase looks very much like a pellet. Pour off the top phase that contains the protein. Its OD_{280}/OD_{260} ratio should be 0.9 ± 0.1. Presumably it is not higher due to remaining oligo- and mononucleotides and other small molecules.
7. At this point, you have a crude protein mixture containing 6.9% PEG. Depending on the use of the crude protein, the PEG can be left in or removed. If the extract is to be put on an ion exchange column in the next step, then excess NaCl can be removed by dialysing against 10–40 vols of buffer for 24 h or diluting with buffer to an appropriate ionic strength. At this point the extract can be loaded directly on most ion exchange columns because the PEG will not interfere with the chromatography.
8. If the extract must be concentrated for gel filtration or if an ammonium sulfate cut is desired, the PEG must be removed. This is accomplished by adding $(NH_4)_2SO_4$ to 30–35% of saturation (100% saturation = 70.5 g/100 ml). If solid $(NH_4)_2SO_4$ is used, it should be added slowly while stirring the extract. If a saturated $(NH_4)_2SO_4$ solution is used, it can be dumped in. The $(NH_4)_2SO_4$ forms another two-phase system and the solution becomes cloudy. After a spin of 10,000 × g for 1 h, the PEG forms a small oily top phase. Two phases appear at about 24% of saturation in $(NH_4)_2SO_4$, but almost all the PEG is removed by using the higher concentrations of $(NH_4)_2SO_4$.

9. Remove the lower phase with a large syringe fitted with a polyethylene tube. Next, precipitate the protein (p. 62) from this lower phase by adding solid $(NH_4)_2SO_4$ either to 65% saturation or to some lower percentage. Stir the solution in the cold for at least 20 min at 4 °C. Then spin very hard to remove the precipitate. It is usually best to use a Spinco 30 head at 25,000–30,000 rpm (75,000 $\times$ *g*) for 20 min. After redissolving in a small volume of buffer, the resulting crude protein solution should have an OD_{280}/OD_{260} ratio of 1.0–1.2, indicating that less than 3% of the absorbing material is nucleic acid.

Purification of Protein-Bound DNA or RNA

The phasing system described above is based on the observation that in 4 M NaCl almost all proteins partition into the upper phase and all nucleic acids into the lower phase. If a protein binds tightly to nucleic acid, and if this binding is salt-dependent, it is often possible to use somewhat lower NaCl concentrations to transfer most proteins to the top phase while leaving the nucleic acid-bound proteins in the bottom phase. The bound protein can then be transferred to a fresh top phase by raising the NaCl concentration, thereby releasing the bound protein. A 10-fold purification can be obtained for RNA polymerase or Qβ replicase using this procedure.

PROCEDURE

1. Follow the procedures detailed above, but use the no-salt polymer concentrate (recipe B) and do not add NaCl to the extract.
2. After the first low-speed spin, remove the top phase, leaving the lower phase in the centrifuge tube. Make the lower phase 1 M in NaCl and then add to it about 5 vols (the amount is not crucial) of 6.9% PEG dissolved in buffer plus 1 M NaCl, the fresh upper phase. Mix for 1 h and once again separate the phases by centrifugation. This procedure can then be repeated with a 2 M NaCl upper phase and finally with 4 M NaCl in the upper phase. RNA polymerase is found in the final 4 M NaCl top phase, whereas Qβ replicase appears in the 2 M NaCl top phase.

COMMENTS

1. PEG interferes with the Lowry protein assay by forming a precipitate. It does not interfere with OD measurements.
2. PEG also precipitates in TCA and, given enough time, will form a gum on the bottom of the test tube. Thus, it is necessary to filter precipitation assays quickly to avoid counts being trapped.
3. PEG also inhibits some enzymes and stimulates others, so

be very careful to check for anomalies before relying on assays done in the presence of PEG.

4. In the multistep phase partition procedure, it is often productive to do two extractions with fresh top phase of the salt concentration that transfers the enzyme to the top phase. The enzyme may not be completely extracted the first time.

COLUMNS, FRACTION COLLECTORS, AND PLUMBING

The Econo-Columns from Bio-Rad Laboratories are versatile, convenient, and inexpensive glass columns for chromatography. Except for particularly short, fat columns, virtually all required dimensions and volumes are available. These columns are equipped with Luer fittings, a standardized diameter and taper fitting which is also used on hypodermic syringes and needles. Thus, tubing and valves are simple to connect if they also have Luer fittings. The simplest and least expensive column connections are made with 20-gauge hypodermic needles. Intramedic polyethylene tubing PE 60 just fits over these needles and makes a good watertight connection. If these needles are used much trouble may be avoided by cutting the point off the needles and filing down the cut surfaces. If this is done, the needles will not puncture the tubing. If extremely large columns are being run, it is better to use the much larger tubing which comes in 2-ft lengths with Luer connectors at each end (#5852 Adapter, Luer; from Ace Glass Inc., Vineland, N.J.).

The common problem of having a column run dry because the reservoir feeding it runs dry can be easily avoided by arranging the tubing to take advantage of the siphon effect. At some point in its path, the tube leading from the reservoir to the top of the column should be lower than the effluent tube end that drips liquid into the collection tubes (Figure 4.1). If all connections are airtight, the top surface of the liquid feeding the column will not go beyond the point which is at the same level as the dripping end. This arrangement allows passage of a specific volume through an unattended column with no danger of having the column run dry.

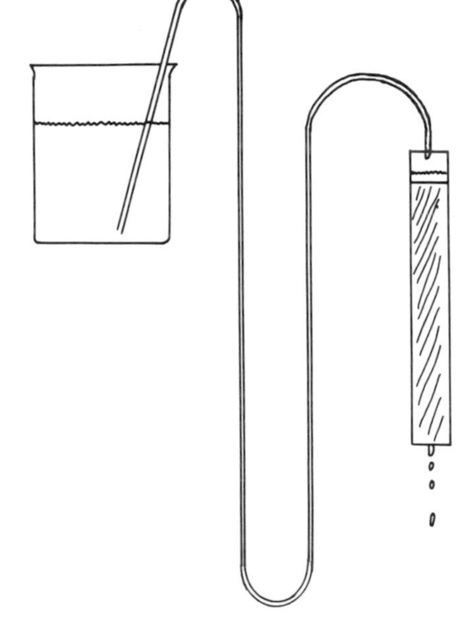

Figure 4.1. A hydrostatic method to prevent columns from running dry.

When using a fraction collector to collect important fractions from a column, it is safest to assume that if anything can go wrong, it will. The Gilson Micro Fractionator is our choice among the currently available collectors. In addition to being small, versatile and reliable, it is constructed to permit recovery of a sample if the collection tubes overfill. One precaution not sufficiently emphasized in the directions for these collectors is that the trolley rods for the collector head must be kept clean. They should be wiped off with a damp towel to remove dried salt, sucrose, and the like. The collector head will advance erratically if the rods are dirty. After about a year of use, these fraction collectors often become unreliable by not indexing to the next collection tube. This results from wear on a guidance rod as it follows a template located under the plastic platform of the collector. Greasing the rod provides temporary repair, and permanent repair is accomplished by having a small ball bearing (B2-37-S, W. M. Berg Inc., East Rockaway, N.Y.) attached to this rod.

ION EXCHANGE CHROMATOGRAPHY AND GEL FILTRATION

The information contained herein should be used in conjunction with the essential information contained in manufacturers' literature: the Pharmacia booklets, *Sephadex Ion Exchangers, a Guide to Ion Exchange Chromatography* and *Gel Filtration in Theory and Practice;* the Whatman publication, *Advanced Ion-Exchange Celluloses Laboratory Manual;* and the *Bio-Rad Laboratories Price List,* which despite its name contains information on materials, equipment, and systems for chromatography.

We will discuss only the commonly used ion exchange resins, DEAE-cellulose, TEAE-cellulose, and phosphocellulose. Many other ion exchangers, such as the basic, acidic, and mixed-bed ion exchange resins and affinity columns as well as hydroxyapatite, are also used to separate proteins. They may be used with essentially the same methods as are described in this chapter. The applications for and use of filtration gels such as Sephadex and Bio-Gel are well described in the manufacturer's booklets and particularly well in the Pharmacia booklet, *Gel Filtration in Theory and Practice.*

Preparing Ion Exchange Resins

Wash dry forms of DEAE as a slurry in 0.5 M HCl, in water until the pH reaches 4, in 0.5 M NaOH, in water until the pH reaches 9, and finally wash with buffer. Repeat a wash more than once if orange or brown color continues to be eluted from the resin. Use a Büchner funnel for convenient removal of solvents from the resin. After washing, suspend the resin in a 10-fold larger volume of the desired buffer and let it settle for about 10 min. Suck off the supernatant liquid that contains the fine particles. Repeat this removal of fine particles twice or until the resin particles settle uniformly. DE-52, a washed, preswollen microgranular DEAE-cellulose, can be used without washing with acid and base, but fine particles do have to be removed.

Whatman P-11 and other cellulose phosphates are prepared like DEAE except that the base wash precedes the acid wash. Phosphocellulose is approximately 1 M in phosphate and hence is a strong buffer with about 10 times the buffering capacity of DEAE. Therefore, considerably more NaOH, HCl, or buffer must be used and the pH must be monitored. P-11 tends to be very dirty, so several washes are likely to be necessary. Because of the multiple washes required and the multiple rinses necessary to bring the pH of phosphocellulose to a desired value, washing phosphocellulose is usually an all-day proposition. All of these resins should be stored in buffers to which a small amount of toluene, 0.01% sodium azide, or another bacterial and fungal inhibitor has been added.

The pH of ion exchange resins is crucial to reproducible chromatography. Phosphocellulose is particularly troublesome because it titrates very slowly. For this reason, before pouring the resin in a column, it is usually best to adjust its pH with a concentrated buffer and then to equilibrate it with the required low-ionic-strength buffer. The pH of phosphocellulose is highly dependent on ionic strength. Therefore, the titration will have to be repeated in the second buffer. The pH ranges that are usually used for DEAE and phosphocellulose are between 6.5 and 9. Above pH 9.5, the DEAE groups become uncharged, and the resin loses its ion exchange capacity. For a pH above 9.5, use TEAE-cellulose, a quaternary amino ion exchanger with properties similar to DEAE. It is best to use a buffer which is not adsorbed by the resin: for example, use Tris buffers for DEAE and phosphate buffers for phosphocellulose. Divalent cations should not be present in the phosphocellulose buffer since they will tightly bind to the resin.

Pouring Columns

Before pouring a column, degas the resin by aspirating for a few minutes. This is especially important if the resin has been stored at 4 °C and the column is being prepared at room temperature. Without degassing, such a column will acquire air bubbles that will usually cause problems. Select a column of appropriate size – a length-to-width ratio of about 10:1 is often satisfactory. Close off the bottom and add a few milliliters of buffer to wet the bottom. Prepare a slurry of the equilibrated resin in about 3 vols of the desired column buffer. Pour the slurry into the column and let it settle at least several minutes before starting the flow. This procedure will prevent light particles from clogging the bottom disk. After this settling, pour the remainder of the resin bed with the column buffer draining slowly through the resin. It is best to pour the column in one pass, usually by attaching an extension to the top of the column. If this is not easy to do, it is satisfactory to fill the column once with slurry and then to add more as the liquid column decreases. It is important to avoid forming an interface between a packed resin surface and a new packing slurry. After pouring, check that the pH and conductivity of the eluent are the same as those of the column buffer. Usually, it is wise to pass several column volumes of buffer through the column before the column is loaded with a sample.

Be sure to choose a column that has sufficient exchange capacity. About 50 mg of crude *E. coli* proteins per ml of packed resin will not overload a DE-52 column. At least 10 times as much can be applied to phosphocellulose. The theoretical protein-binding capacity of phosphocellulose is 10 times that of DEAE, but the adjacency of phosphate groups may limit this. Only 10% of a mixture of crude *E. coli* proteins binds to phosphocellulose whereas 95% binds to DEAE; hence, in theory one could apply 100 times as much crude protein to a phosphocellulose column as to a DEAE column of the same size. Our experience, however, contradicts theory, so we recommend a lower quantity. To avoid tailing of protein peaks during analytical separations, it is best to use about 10% of these amounts, i.e., 5 mg/ml for DEAE and 50 mg/ml for phosphocellulose. Don't go overboard. A column that is too large will result in unnecessary dilution of protein. Such dilution can be more than an inconvenience because protein concentrations lower than about 0.1 mg/ml will reduce protein stability.

Loading and Eluting Columns

Apply the sample to the column without disturbing the column surface. DEAE may be loaded under pressure because adsorption is so rapid that it occurs in minutes. Phosphocellulose appears to bind proteins more slowly than DEAE. Hence, it is preferable to allow the maximum convenient time for passing the loading solution through a phosphocellulose column. Rinse the loaded column with sample loading buffer until no protein is found in the eluent. This usually takes about three column volumes. Sufficient rinsing is especially important for phosphocellulose since most of the protein being passed through the column will not bind even a low ionic strength (see below).

For maximal separation between proteins it is best to elute them from the resin with a continuous salt gradient, which can be formed as shown in Figure 4.2. Step changes in salt concentration applied to a column are usually most valuable for concentrating a protein solution or for removing bulk impurities such as nucleic acids (nucleic acids are separated from most proteins by eluting the proteins from DEAE with 0.3 M KCl). The gradient can be about 10 column volumes. Much larger gradients are sometimes used, but probably do not increase resolution and certainly do lead to greater dilution of the proteins. Arrange the gradient apparatus so that the buffer volume at the top of the column is small but suf-

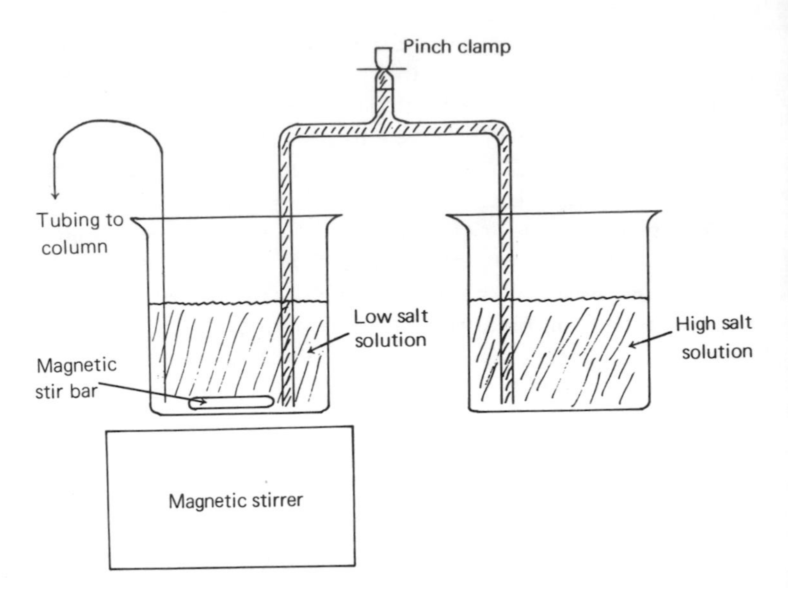

Figure 4.2. One method for producing a gradient in salt concentration for the elution of proteins from a column. After loading each half of the apparatus with the desired buffer, the U tube is filled with liquid by sucking on the middle T joint with a pipette and then sealing this tube. Note that the low salt buffer is loaded into the reservoir that is connected directly to the column. The final buffer to emerge from the apparatus is equal to the buffer loaded into the second reservoir. The U tube should be narrow enough to prevent macro mixing: that is, <3 mm diameter.

ficient to prevent the inflowing liquid from disturbing the surface of the resin. DEAE and phosphocellulose columns can be operated under pressure. A convenient flow rate can usually be obtained with a hydrostatic pressure of 0.5–1 m. In general, the slower the flow rate, the better the resolution.

To elute proteins from DEAE, KCl gradients are usually used. Of a mixture of *E. coli* proteins, approximately 5–10% flow through DEAE at 0.05 M KCl, and 5–10% still bind at 0.30 M KCl. The highest protein concentration elutes at about 0.15 M KCl. To elute proteins bound to phosphocellulose, phosphate gradients are frequently used, although KCl may be employed as well. Phosphate gradients have the advantage of providing increasing buffering power at higher ionic strengths. This may be important because the resin gives off protons as the ionic strength increases and, as previously mentioned, pH has a significant effect on a protein's adsorption to the resin. Most of the proton release occurs in the range below 0.15 M phosphate. In a phosphocellulose column being run at pH 6.5 with a gradient of from 0.02 to 0.50 M phosphate, the pH drops to a low of 5.8 at about 0.15 M phosphate. In a similar gradient of from 0.10 to 0.50 M, the pH drop is less than 0.2 pH units. Thus if the protein of interest binds at a high salt concentration, gradients should not be started at very low ionic strengths.

After running the column it is informative to learn the salt concentration in the various fractions. This is particularly useful for smaller columns in which it is difficult to control development of the salt gradient. Since conductivity is nearly proportional to ionic strength and is easily quantitated, it is the measurement of choice. Calibrate a conductivity meter with samples of known salt concentration in the column buffer. Be aware that conductivity is strongly dependent upon temperature and that a sample's conductivity can change appreciably between 4 and 0 °C. Therefore, it is best to have all samples in ice water. Conductivity meters cover a large range and are highly sensitive, so that it may be more convenient to dilute a small expendable portion of each fraction with distilled water and to make the measurement on this.

The elution properties of protein may change in the presence of nucleic acids. Nucleic acids generally bind more tightly to DEAE than do proteins and hence proteins tightly stuck to DNA, for instance, may elute from a DEAE column at very high ionic strengths. Examples for this phenomenon are T4 ligase and RNA polymerase. The RNA polymerase activity of an extract from which most nucleic acids have been removed elutes in a single peak from DEAE at 0.17 M KCl. Without prior nucleic acid removal the activity peaks at 0.17 M and again at 0.35 M KCl, with activity spread throughout the gradient from 0.17 to 0.40 M KCl. On phosphocellulose, the opposite effect may occur, that is, proteins bound to nucleic acids may flow through the column under conditions where they might otherwise bind. Most proteins probably are not bound to nucleic acids, but even in such cases activity peaks on DEAE tend to be broader if crude solutions are applied to the resin. Proteins can be separated from nucleic acids by dextran–polyethylene glycol phase partition, protamine sulfate precipitation, or 0–50% precipitation with ammonium sulfate. If the protein of interest binds to phosphocellulose and not to DNA, a small phosphocellulose column will be effective as a first purification step.

Recently, a convenient method has been developed for removing nucleic acid during purification of *E. coli* RNA polymerase and cyclic AMP receptor protein (Burgess and Jendrisak, 1975; Eilen et al., 1978). In this method the polycation polyethylenimine (Polymin P, obtainable from Sigma Co.) precipitates DNA but few proteins. In some cases a protein precipitated by Polymin P can be released from the precipitated DNA by increasing the salt concentration in the buffer. Insufficient experience has been acquired to assess the general utility of Polymin P as an early step in purification of other proteins.

After the gradient is finished, the column can be reused if it is washed in a highly concentrated salt solution, for example, 1.0 M KCl or phosphate. If the top surface still looks discolored it should be removed. If a crude extract labelled with ^{32}P is applied to DEAE, some counts still stick after a wash; hence, these washes do not totally regenerate the columns. If only proteins that are fairly free of nucleic acid are applied to columns, the columns may be reused for at least 10–20 times. The flow rate may drop after the first high salt wash because the resin contracts at higher ionic strength and is unable to expand to the original volume at low salt. To minimize the flow rate decrease, the top half of the resin can be stirred with a Pasteur pipette and then allowed to resettle in a low-ionic-strength buffer.

DETERMINING PROTEIN CONCENTRATION

Optical Density

One milligram of an average protein in 1 ml in a 1-cm cuvette gives an OD_{280} of 0.55. The absorbance at this wavelength is due to tryptophan and tyrosine so that this value can vary by 50% from one protein to another.

Whenever nucleic acid is present, a correction must be made for its contribution to OD_{280}. One such approximate correction is as follows:

$$\text{Protein concentration (mg/ml)} = 1.5 \times OD_{280} - 0.75 \times OD_{260}.$$

More sensitive measurements of low protein concentrations are sometimes possible by using a shorter wavelength of light. For example:

$$\text{Protein concentration } (\mu\text{g/ml}) = 144\,(OD_{215} - OD_{225}).$$

Note that β-mercaptoethanol and dithiothreitol both absorb strongly in the wavelengths below 280 nm and that their absorbance changes as they oxidize. EDTA also absorbs strongly in the short-wavelength UV region. Therefore, when measuring protein concentrations by UV absorbance, for example, after elution of proteins from a column, it is best to use a buffer blank of the same age and composition as that containing the sample.

Biuret

This is a rather insensitive assay but its speed and simplicity make it useful for some applications. Note that NH_4^+ interferes with this assay, so use appropriate controls and standards. Mix 1 ml of the biuret reagent with up to 100 μl of sample and let the mixture sit for 20 min at room temperature. Read OD_{550} against the reagent plus buffer. An OD of 0.06 corresponds to about 0.2 mg.

BIURET REAGENT

1. Make 100 ml of 0.2 N NaOH, 0.8 g NaOH/100 ml distilled H_2O. Boiling the H_2O before adding NaOH pellets appears to prolong the lifetime of this solution.
2. Dissolve 2.25 g sodium potassium tartrate in the NaOH solution.
3. Add 0.75 g $CuSO_4 \cdot 5H_2O$. To assist dissolving, grind the crystals in a mortar before adding them to the solution.
4. Add 1.25 g KI and dissolve completely.
5. Transfer all of this to a graduated cylinder and add dH_2O to a total volume of 250 ml.

6. Store in a polyethylene bottle because glass reacts with the base.

Lowry

This assay is sensitive and reproducible. A calibration curve with a protein such as BSA must be determined because the response is nonlinear. The assay responds to tyrosine and tryptophan, so different proteins give different amounts of color when equal weights are used.

RECIPES

2% (wt/vol) $CuSO_4$
4% (wt/vol) K^+ tartrate
3% (wt/vol) Na_2CO_3 in 0.1 M NaOH
Phenol reagent (Folin – Ciocalteau)

On the day of the assay prepare:
(1) 2 ml $CuSO_4$ solution plus 2 ml of K^+ tartrate plus 98 ml of the Na_2CO_3 solution.
(2) Dilute the phenol reagent with an equal volume of H_2O.

PROCEDURE

1. Adjust aliquots containing 10–100 μg of protein to a volume of 0.5 ml.
2. Add 5.0 ml of the $CuSO_4$-tartrate-Na_2CO_3 mixture.
3. After 10 min at room temperature add 0.5 ml of diluted phenol reagent and shake vigorously for 10 sec.
4. Read the OD_{650} after an additional 30 min.

Lowry for Dilute Samples

This procedure can measure as little as 1 μg of protein in about 0.8 ml of solution. This sensitivity requires extreme care, however, and probably triplicate samples are necessary to be sure of such a value. A good range to measure with this modification is 2–100 μg of protein. The sample can be in any buffer and thiols do not interfere as they do in the unmodified Lowry procedure. The buffer may contribute a small background, however, and for the most careful work a buffer blank should be run.

RECIPES

Reagent Composition

A. 2% Na_2CO_3 (wt/vol)
B. 0.5% $CuSO_4 \cdot 5H_2O$ (wt/vol), plus 1% Na^+-citrate

C. A + B, 50:1 (made fresh daily)
D. Phenol reagent (Folin–Ciocalteau), diluted 1:2 with water just before use

PROCEDURE

1. Add the required amount of sample to a 1.5-ml Eppendorf microfuge tube and dilute to 0.8 ml with water or buffer. Chill the sample for a few minutes in ice water.
2. Add 0.2 ml of ice-cold 50% (wt/vol) trichloroacetic acid and allow the mixture to sit at least 30 min on ice. Then centrifuge in an Eppendorf microfuge for 2 min and pour off the supernatant away from the outer side of the tube that has the invisible pellet.
3. Add 0.2 ml of 0.5 M NaOH, vortex, and allow to stand at room temperature for more than 2 h.
4. Add 1 ml of reagent C, vortex, and allow to stand 10 min; then add 0.1 ml phenol reagent (D) and allow to stand 30–120 min at room temperature.
5. Read the absorbance at 660 nm.

A standard curve must be run with each batch of samples. A reasonable standard curve consists of 12–15 points, for example, 0, 1, 2, 5, 10, 20, 25, 50, 75, and 100 μg of BSA in H_2O. If the curve is to be precisely defined at the lower end, duplicates are essential.

CONCENTRATING PROTEIN SOLUTIONS

Many purification schemes leave the protein at a concentration too low to be convenient for storage or too low for the subsequent experiments. The safest way to concentrate a precious enzyme is to place it in a dialysis sack and embed it in Sephadex G-100 or G-200 at 0–4 °C. The buffer is able to pass through the pores in the membrane but the protein molecules cannot. As the buffer comes through the membrane, it is absorbed by the Sephadex and more buffer can pass through. To concentrate 3 ml down to a volume of 0.5 ml, the gummy layer of hydrated Sephadex should be peeled off the sack once every half-hour for about 3 h. The same method will work with materials less expensive than Sephadex, for example polyethylene glycol, but they occasionally leach OD absorbing material (and possibly other substances) into the sack. Dialysis against buffers containing 5% polyethylene glycol will also concentrate protein solutions.

A variety of companies sell equipment which is convenient for concentration of protein solutions. The pieces of equipment sold by both Amicon and Millipore Companies work well and are sensible to use if concentration is a step which must be performed often.

A protein solution can be concentrated by precipitating with ammonium sulfate, resuspending in a smaller volume, and removing the ammonium sulfate by dialysis or passage through a Sephadex G-25 column. If you are deeply involved in protein purification, it pays to have a set of Sephadex G-25 columns with different volumes. These can be calibrated so that the excluded and included volumes can be collected and pooled on the basis of volume alone, and so that they allow a complete separation of the excluded and included volumes for the particular sample volume applied to the column. By using a column with a volume 3–4 times the volume of your sample, you can change buffers or remove a salt in half an hour with no more than a 50% increase in sample volume.

Occasionally, a sample can be concentrated by binding it to an ion exchange column and eluting in a smaller volume.

Vacuum dialysis is a very gentle and fairly rapid concentration method. Not only is it inexpensive, but it also can be used to handle many samples or to concentrate an unlimited volume to any desired smaller volume. For handling many samples, the bottom of a desiccator jar is used. A 0.5-in. piece of Lucite is bored with as many 0.5-in. holes as needed. Each of these is used as shown in Figure 4.3. It should be noted that during concentration it is possible to change the buffer by bathing the outside of the dialysis sack with a new buffer.

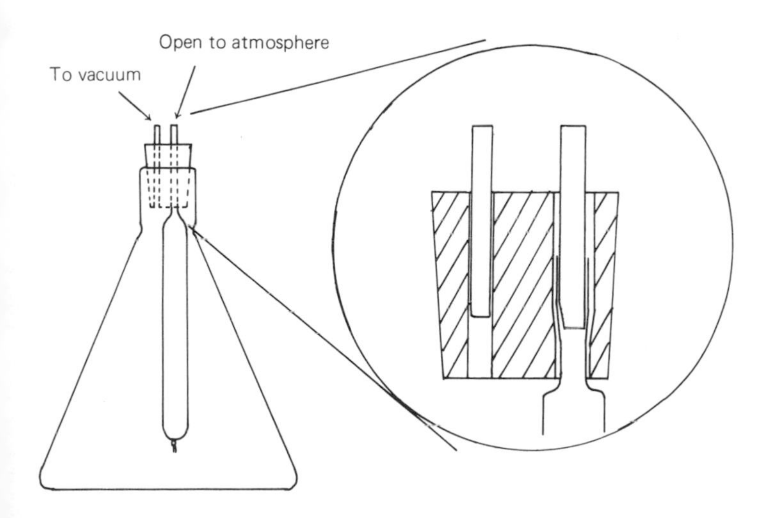

Figure 4.3. An apparatus for vacuum dialysis. Note that the flask can contain a buffer different from that of the sample being concentrated. Thus, a sample may be concentrated and at the same time have its buffer changed.

The sack is attached to the system by pushing it through the stopper. Then the tapered end of a truncated Pasteur pipette is pushed into the sack, and a seal is formed by squeezing the sack against the sides of the hole in the stopper.

STABILIZING PROTEINS

It usually happens that the protein central to your study is extremely labile. Although it is impossible to predict what steps will stabilize an uncharacterized protein, a number of procedures are frequently helpful and are worth testing. A list of occasionally successful ploys is given below.

STEPS

1. After growing cells and before freezing, wash them 3 times in 0.05 M Tris, pH 7.5, 10^{-2} M Mg^{2+}-acetate, 10^{-3} M EDTA, 10% glycerol, and an SH reagent, for example 2-mercaptoethanol (see step 2).
2. Any of the following SH reagents can be used to help stabilize a protein: 10^{-4} M Cleland's (dithiothreitol), 10^{-3} M glutathione, or 5×10^{-4} M 2-mercaptoethanol. Cleland's is best, but expensive. Fewer proteins are sensitive to it than to mercaptoethanol. Most proteins are stabilized by an SH reagent, but a few are sensitive to such reagents.
3. If it is necessary to store cells before purifying the protein do so at −70 to −120 °C rather than at −20 °C.
4. Glass beads may be inappropriate for opening cells because glass adsorbs some proteins. For labile proteins, it is better to open cells by grinding in alumina (p. 15).
5. All plastic ware should be boiled 10 min in 10^{-3} M EDTA. All glassware should be swirled with dichromate cleaning solution, rinsed with 0.01 M sodium citrate, and thoroughly rinsed with water. Be aware that such glassware seems unusually fragile.
6. During the opening of cells, it is best to have 0.1 M KCl in the opening buffer.
7. Bear in mind that some proteins stick to dialysis tubing.
8. Any dialysis tubing to be used should be boiled (p. 186).
9. During ammonium sulfate precipitation, add it slowly and hold the pH constant with KOH. Use enzyme grade ammonium sulfate. To remove ammonium sulfate after precipitation, dialyse 3 times for 1.5 h and check for 10^4 removal of ammonium by conductivity (p. 72).
10. Proteins that cannot stand freezing can be stored in 50% glycerol at −20 °C.
11. A few proteins are killed by O_2 and all work with them must be done under argon.
12. Molecules which bind to a protein frequently stabilize it. Substrates, products, or inhibitors of enzymes are often good examples.

POLYACRYLAMIDE GEL ELECTROPHORESIS OF PROTEINS

A convenient method for quantitating and characterizing a protein is electrophoresis in an acrylamide gel. Electrophoresis procedures are rapid and detecting the proteins in the gel by staining or autoradiography is convenient and sensitive. Hence, it is simple to perform electrophoretic separations on large numbers of samples so that, for example, electrophoresis can be used to assay fractions eluted from an ion exchange column.

Electrophoretic migration rates have a fairly predictable relation to the molecular weight of proteins if the proteins are dissociated and denatured with sodium dodecyl sulfate (SDS) before and during electrophoresis (Weber and Osborn, 1969). Most proteins bind a fixed amount of SDS per amino acid; as a result, proteins acquire a fixed charge-to-mass ratio and differ only in their size. Electrophoresis of such proteins then separates them strictly (almost) according to size.

The utility of acrylamide gel electrophoretic separations of protein is so great that a wide variety of modifications have evolved from the basic technique. The major classes of these modifications are as follows:

1. Straight SDS electrophoresis.
2. SDS electrophoresis with a stacking gel. A stacking gel is a porous, low percentage gel on top of the separation gel. It permits samples of large volume to be electrophoretically concentrated into a thin band during the initial stages of electrophoresis. Sharper protein bands result.
3. Gradient gels. By adjusting the concentration of acrylamide and/or cross-linker, gels can be optimized for the separation of proteins in a particular molecular weight range. The use of appropriate gradients extends the range over which any given gel will produce good separation.
4. Buffer gels. Instead of electrophoresis of denatured, SDS-treated proteins, these gels allow nondenatured proteins to undergo electrophoresis.
5. Urea gels. These semidenaturing gels also have been known to provide useful information.
6. Isoelectric focusing gels. These gels separate proteins according to their isoelectric point.
7. Two-dimensional gels. These gels are of extraordinarily great value. Proteins are first separated according to their isoelectric point, and then, in the second dimension, they are separated according to their molecular weight. Use of these gels is fully described in O'Farrell (1975) and O'Farrell et al. (1977).
8. Radioactive gels. Localization of radioactive proteins in gels can be performed by cutting out and counting pieces of the gel, but fluorography of the gels is more convenient and usually gives information of higher resolution. By this method extremely small quantities of protein may be localized and quantitated in any of the types of gels mentioned above.

In this section we will describe protocols for 10% SDS–acrylamide gels, exponential gradient gels, and urea–SDS gels and for the staining, drying, and fluorography of gels.

Making a Gel Sandwich

Make a gel sandwich from two glass plates that are 14 cm wide, 21.5 cm high, and 3 mm thick by separating them with Lucite spacers that are 1–2 mm thick and 1.4 cm wide (Figure 4.4). Experience with combs and spacers leads us to recommend that all spacers and combs be the same thickness within 0.002 in. Often Lucite stock is not this flat and must be fly cut to adequate flatness. A notch in the rear glass plate is 11 cm wide and 2 cm high. The ears which form the edge of the slot are easily broken and plates with this notch are more expensive. For these reasons it makes more sense to use a plate as shown. Portable ears may then be added to the sandwich and held in place with Dow Corning Vacuum Grease. Add the ears after inserting the sandwich in the reservoir apparatus and before adding buffer. Place the spacers on the two sides and the bottom of the sandwich. Remove the spacer that goes on the bottom after the gel has polymerized.

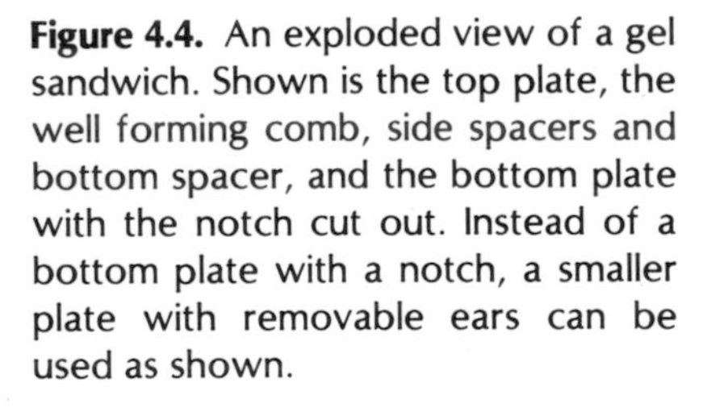

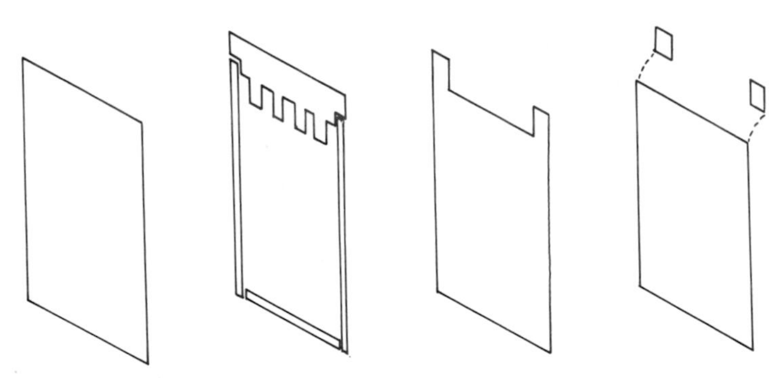

Figure 4.4. An exploded view of a gel sandwich. Shown is the top plate, the well forming comb, side spacers and bottom spacer, and the bottom plate with the notch cut out. Instead of a bottom plate with a notch, a smaller plate with removable ears can be used as shown.

Two useful methods exist for sealing the sandwich. One method uses a thin layer of vacuum grease as a sealer. Put the grease first onto the bottom plate, add the spacers, then grease the exposed surfaces of the spacers, and add the top plate. Use a very thin trail of grease and add it by extrusion through a hypodermic syringe. After assembly, squeeze the sandwich and hold it between large binder clips (Figure 4.5). Minimize the grease which enters the space where the gel will be formed, as grease retards gel polymerization. After electrophoresis remove the grease from gel plates and spacers by washing them in ethanol. Thoroughly wash the plates, since grease on a plate can also cause protein bands to smear.

The second method of sealing the sandwich uses paraffin (canning wax, Gulf Oil Corporation, available at hardware stores) and requires a pan of molten paraffin. Heat the pan containing paraffin in a pan of water for about 30 min on a hot plate. In this method, hold the sandwich together with 1 binder clip on each side. Dip the bottom edge and then the bottom corners about 2 cm into the molten wax. Allow

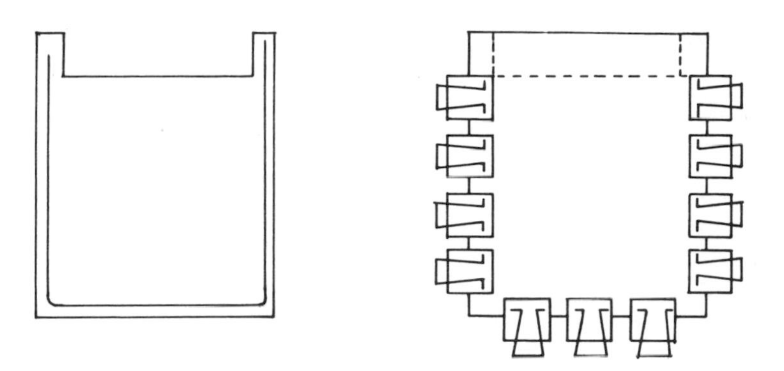

Figure 4.5. The area of the gel plate which must be greased in order to seal the sandwich is shown by the U-shaped line on the notched plate. The arrangement for holding the sandwich together with binder clips is also shown.

the sandwich to cool about 30 sec, then remove 1 side clip and dip that side. Then remove the other clip, and dip that side. After dipping, allow the sandwich to cool about 5 min, during which time cooling cracks will form in the paraffin. If any cracks are visible in the paraffin seal, redip the sandwich and tilt it so that the cracks are filled. These gel sandwiches rarely leak. After using the gel, remove the wax by scraping. The wax may be reused. Remove the wax residue in the plate by washing in warm water. Once the sandwich has been sealed, make the gel solution and pour.

SDS–10% Acrylamide Gel with Stacking Gel

In our experience, there is very little concerning this recipe that is critical. The dimensions, acrylamide concentrations, number of teeth in the comb, and the presence or absence of a stacking gel can all be changed. Changing any of these variables does have an effect, however, so it is sometimes useful to systematically search in order to find the optimum conditions for the particular application. The technique is basically that first described by Laemmli (1970).

RECIPES

1. *Ingredients of 10% Gel*

10 ml 30% (wt/vol) acrylamide (Bio-Rad), 0.8% *N,N'*-methylenebisacrylamide (Eastman or Bio-Rad) in dH_2O. Acrylamide stock solutions can be stored for months in a refrigerator
3.75 ml 3 M Tris–HCl, pH 8.8
15.8 ml distilled H_2O
(degas)
300 μl 10% SDS (Most lots of most brands work but occasional lots will precipitate in the gel. The sequanol grade SDS from Pierce Chemical Co. has been consistently satisfactory.)
15 μl *N,N,N',N'*-tetramethylene-ethylenediamine (TEMED; Eastman or Bio-Rad)
150 μl 10% (wt/vol) freshly dissolved ammonium persulfate (Bio-Rad)

2. *Ingredients of a Stacking Gel*

1.25 ml 20% (wt/vol) acrylamide, 0.8% *N,N′*-methylene-bisacrylamide in distilled H_2O
1.25 ml 1 M Tris-HCl, pH 8.8
7.34 ml H_2O
(degas)
100 μl 10% SDS
6 μl TEMED
60 μl 10% ammonium persulfate (Bio-Rad)

3. *2× Sample Buffer*

0.15 g Trizma base
0.4 g SDS
1.0 ml 2-mercaptoethanol (98%; Eastman Chemicals)
2.0 ml glycerol
7.0 ml distilled H_2O
0.02 g bromphenol blue
Store at −20 °C in 100 μl aliquots

4. *Electrophoresis Buffer*

14.4 g glycine
3.0 g Trizma base
10 ml 10% SDS
800 ml distilled H_2O
Adjust pH to 8.3 if necessary, and bring to a final volume of 1 liter

PROCEDURE

1. One preliminary warning is necessary. Acrylamide is reputed to be a toxin of the central nervous system.
2. To prepare a 10% gel, first mix the acrylamide, Tris, and water in a 125-ml side-arm filtration flask. Connect the flask to a vacuum aspirator, stopper the top, and suck. Continue the sucking with gentle shaking for several minutes or until all of the dissolved gas has been removed. This step is necessary to remove dissolved oxygen which would otherwise poison the acrylamide polymerization reaction. After the degassing step, add the final three ingredients. Following these additions, it is best to work rapidly, for polymerization has begun.
3. Immediately after making the 10% gel solution, pipette it into the vertical sandwich and fill to 3.5 cm from the well notch at the top back of the gel. Then gently layer 1 cm of H_2O on top of the acrylamide solution without disturbing the acrylamide. Usually it is easiest to add the water at the very edge of the sandwich. It is informative to put 2–4 ml of gel solution in a Pasteur pipette in order to test polymerization of the gel. The remainder of the gel in the flask may not polymerize. Allow at least 1 h for the gel to polymerize. There should be a clearly visible interface

between the gel and the water and the gel in the pipette should be hard. When the 10% gel has hardened, make the stacking gel solution. Pour the water off the top of the 10% gel. Ignoring the small amount of water left on top of the 10% gel, immediately pour the stacking gel solution on top of the 10% gel. Put a well-forming comb in place and let the stacking gel polymerize for at least 0.5 h. Remove the clamp and flush out the wells to remove unpolymerized acrylamide. Leaving the comb in the gel for more than 6 h increases the chances that the wells will dry out. The gel may be stored several days at room temperature if the wells are filled with electrophoresis buffer.

4. After polymerization, remove the bottom spacer and clamp the gel sandwich in the reservoir apparatus as shown in Figure 4.6. Two problems occur infrequently, and are best dealt with at this point in the procedure. First, leaks between sample wells can occur. If the well partitions look suspect, test them by filling every other well with electrophoresis buffer. If there is leakage, it can be stopped by squeezing the gel sandwich. It can be squeezed by placing a toothpick behind the middle of the rear plate between the plate and the reservoir. Stubborn cases require the use of a large ringstand clamp pressing

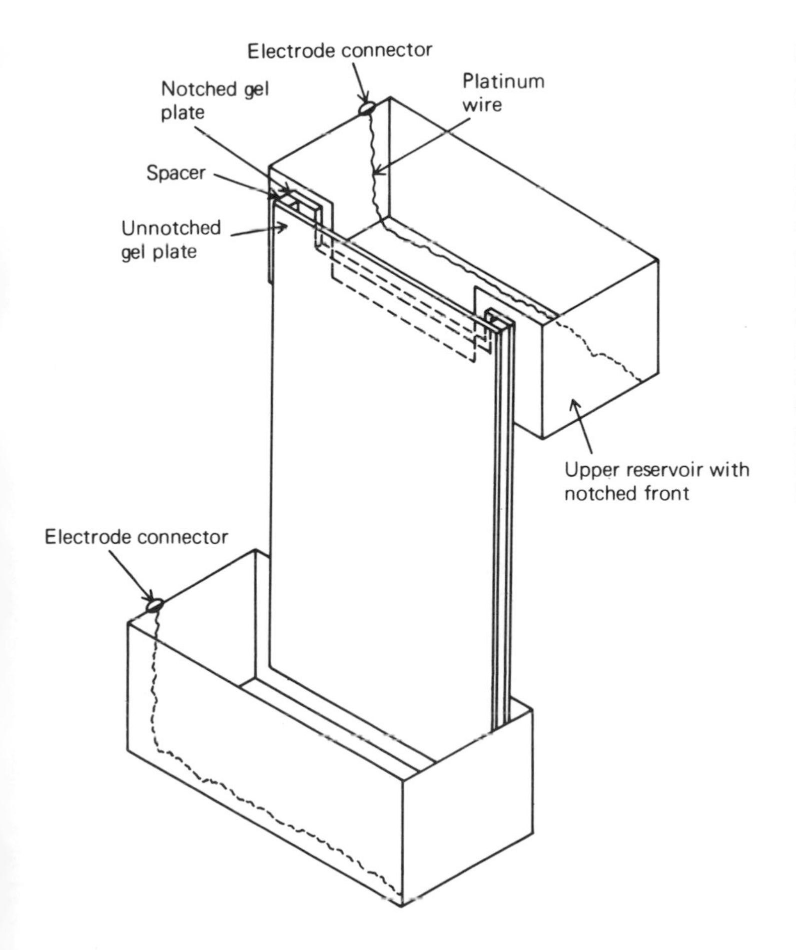

Figure 4.6. The vertical gel electrophoresis apparatus with reservoirs. The apparatus is Plexiglàs. The upper reservoir is attached to a vertical metal rod. A thumbscrew allows it to be moved to accommodate gel sandwiches of different sizes. A trough in the upper reservoir (dashed line) is filled with 0.5-cm diameter rubber tubing and forces a seal between the reservoir and the gel sandwich. The upper reservoir is filled high enough that liquid makes direct contact with the gel. The bottom reservoir is filled high enough to make contact with the bottom of the gel.

on the middle of the front plate and spanning the distance between the front plate and the rear of the upper reservoir. The second problem occurs only with grease-sealed sandwiches. The grease can interfere with acrylamide polymerization and yield unpolymerized channels along the edge of a side spacer. As a result, liquid will pass from the upper to the lower buffer tank. It can be cured by removing clamps on the offending side and pushing the side spacer in with a spatula. After these steps, complete filling the reservoirs with electrophoresis buffer so that it makes contact with the gel at the top and bottom. Remove any trapped bubbles at the bottom with a bent-tipped Pasteur pipette.

5. Prepare samples for electrophoresis by boiling for 2 min in 1X sample buffer. Usually 0.25 μg of protein in a single band is visible and more than 20 μg in a band or 100 μg in a well overloads that lane. Up to 100 μl can be loaded in a single well. The glycerol in the sample buffer permits layering of the sample directly on top of the gel and under the electrophoresis buffer. A mechanical pipetting device (Clay Adams) connected to a 100-μl disposable pipette facilitates loading the sample without unduly stirring it up with extraneous bubbles. As you load a well, place the tip of the pipette just above the gel.
6. Electrophoresis in these gels is usually for about 3 h at 60 mA, which generally yields a voltage of between 120 and 200 V. When crude extracts of *E. coli* are undergoing electrophoresis, and probably in most other situations as well, the sharpest bands result when the current is as high as possible without cracking the glass plates from heat. Optimally, the plates are hot but do not burn if they are touched. Generally, electrophoresis is continued until the bromphenol blue dye is about 1 cm from the bottom of the gel.

Urea–SDS–Acrylamide Gradient Gels

This gradient gel was developed by Storti et al. (1976). The version we describe is an 8–12% exponential gradient that gives excellent resolution of proteins in the 55,000-dalton range when the bromphenol blue tracking dye has reached the bottom of the gel. The gel percentages can, of course, be altered to increase the resolution in other molecular weight ranges.

RECIPES

Solution A

30.0 g acrylamide (Bio-Rad)
0.8 g *N,N'*-methylenebisacrylamide (Eastman or Bio-Rad)
30.0 g urea (Schwarz/Mann; ultra pure)

Dissolve and bring to a final volume of 100 ml with distilled H_2O
Store at 4 °C in an opaque bottle

Solution B

18.17 g Trizma base (Sigma)
40 ml distilled H_2O
Adjust pH to 8.8 with HCl
Add 48.05 g urea (Schwarz/Mann; ultra pure) and bring to a volume of 98 ml with distilled H_2O
Adjust to pH 8.8 with HCl. Bring to a final volume of 100 ml with distilled H_2O. This may be stored at room temperature for many months if the pH is readjusted when necessary.

Solution C

6.06 g Trizma base (Sigma)
30 ml distilled H_2O
Adjust pH to 6.8 with HCl
Add 48.05 g urea (Schwarz/Mann; ultra pure)
Stir and heat to bring the urea into solution (do not heat above 60 °C or the urea will degrade)
Bring volume to 98 ml with distilled H_2O
Adjust pH to 6.8 with HCl and bring to a final volume of 100 ml
Store at room temperature. It is stable for many months but the pH should be readjusted just before use.

Gel Solutions

8%	12%	Ingredient
5.88 ml	2.8 ml	solution A
5.25 ml	1.75 ml	solution B
7.06 ml	2.35 ml	10 M urea
2.52 ml (degas)	—	distilled H_2O
8.4 μl	2.8 μl	TEMED
84 μl	28 μl	ammonium persulfate
210 μl	70 μl	10% SDS

PROCEDURE

1. Make a gel sandwich and gel solutions as described for 10% acrylamide gels. Shut the two valves of the gradient maker (Figure 4.7). Immediately after making the 8% and 12% gel solutions, pour 7 ml of the 12% solution into the chamber with the outlet connected to the gel sandwich and 21 ml of the 8% solution into the other chamber. Drop an 8 × 1.5 mm stirring bar into the chamber with the 12% acrylamide solution and plug the top of that

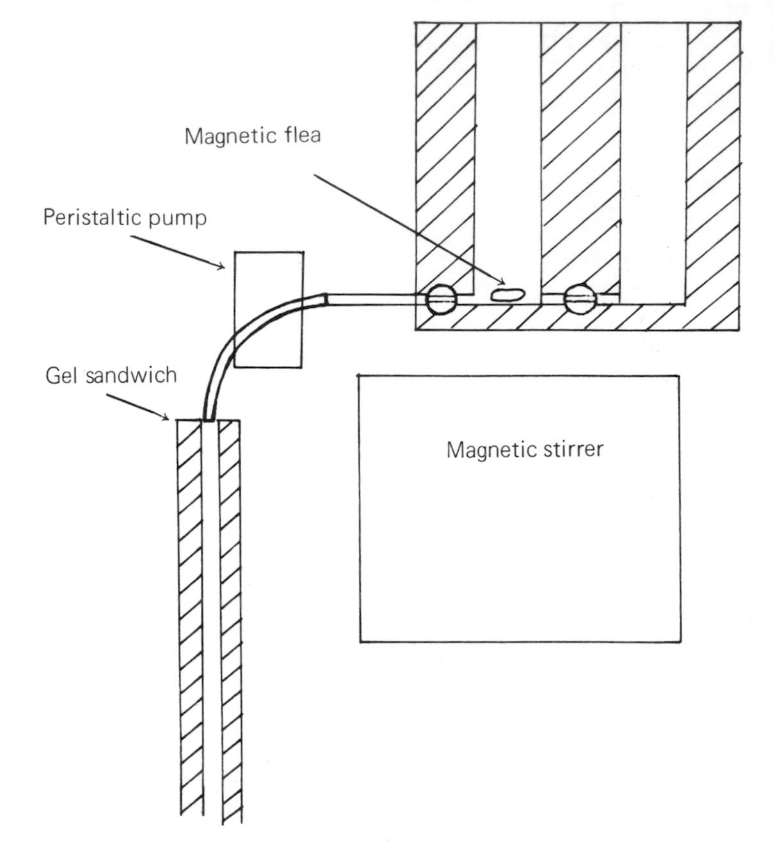

Figure 4.7. An arrangement for making a gradient gel.

chamber with a rubber stopper so that its liquid level will remain constant.

2. Turn on the magnetic stirrer so that it gently mixes the solution in the 12% chamber. Open the valve between the two chambers and then the valve to the gel box. Turn on the peristaltic pump to drive the gel solution into the gel box. The tubing leading to the gel box should be taped to the top of the gel plates at the midpoint of the large notch at the back. A setting of 6 (10 ×) on an LKB pump works well. The flow rate should allow the gel formation to be complete in 10–15 min. When the 8% gel solution chamber is empty, close the valve between the two chambers so that no air bubbles are drawn into the other chamber. When all of the acrylamide has entered the gel box, put the end of the tubing into a beaker and flush the chambers and tubing with distilled water so that acrylamide will not harden in the apparatus.
3. Pipette distilled water onto the top of the gel and allow the gel to harden for at least 1 h. Then form a stacking gel on top of this gel by the method described for 10% gels, but using the following recipe. Perform electrophoresis for about 8 h at 120–200 V.

Ingredient	Volume
Solution A	1.0 ml
Solution C	2.5 ml
10 M urea	1.4 ml
Distilled H_2O	4.9 ml
TEMED	5 μl
10% ammonium persulfate	100 μl
10% SDS	100 μl

Staining Gels

The most convenient and sensitive stain for all-around use is Coomassie Brilliant Blue R (Sigma). To stain a gel, first remove the sandwich from the apparatus. Then scrape the edge of the sandwich and pull out the spacer. Lift off the upper plate. If the two plates stick together, pry them apart with a gel spacer or a razor blade. It is best to start prying at the stacking gel portion of the sandwich. The gel will stick to one of the two plates. Detach it from that plate and slide the gel into a Pyrex dish of staining solution A (see below). Float the gel off the glass plate by squirting a layer of water under it with a wash bottle. Stain in solution A for 1–18 h at room temperature. During the staining and destaining cover the tray with Saran Wrap. The staining solution can be used many times.

Times are not critical and any of these steps can continue overnight. Note that the stain, solution A, is reusable. Finally, destain by soaking solution B at least 1 h one or two times. Finally, complete destaining with soakings in 10% acetic acid; typically, 4 h or more are required.

RECIPE

Solution A

450 ml methanol
90 ml acetic acid
2.5 gm Coomassie Blue
450 ml H_2O

Solution B

90 ml acetic acid
450 ml H_2O
450 ml methanol

Fluorography of [^{3}H]- or [^{35}S]-Labelled Proteins in Acrylamide Gels

Stain and destain the gel as previously described. If staining is unnecessary then soak the gel in solution B overnight to fix the proteins and remove unincorporated label. Soak the gel twice for 30 min in 100% dimethyl sulfoxide (DMSO) at room temperature. The first soak can be in 100% DMSO that has been used for the second soak of a previous gel. Then soak the gel for 3 h at room temperature in 22% (wt/vol) 2,5-diphenyloxazole (PPO) in 100% DMSO. *Warning:* always use gloves when working with the reputed carcinogen, PPO. The PPO solution may be used for more than 20 gels. If a large precipitate forms in the stock bottle of the PPO solution, recover the PPO and make a new solution. The gel shrinks during the PPO treatment.

Swell the shrunken gel in distilled water for 1 h and use gloves to wipe PPO precipitate from the surface of the submerged gel. The precipitate will cause autoradiographic artifacts. Flush the surface of the gel with distilled water and then slip a piece of 3-MM paper (Whatman) under it and lift it with the paper. Cover it with Saran Wrap so there are no wrinkles, then put it on a commercial gel dryer (Bio-Rad, for example) and dry it under vacuum for a total of 2 h. Use the heating element of the gel dryer for the first hour. *Warning:* The gel will disintegrate if the vacuum is released before the gel is dry. Use Kodak RP50 film on the side of the gel covered with Saran Wrap and expose it at −70 °C. A linear relationship between radioactive emissions and film exposure can be achieved by preflashing the film (p. 185).

Recovering PPO from Solution in DMSO

Perform this procedure in a hood. Use gloves and lab coat. Protect your face and other skin from any splashes.

Filter the DMSO–PPO solution through Whatman #1 paper using a Büchner funnel. A water aspirator will provide sufficient suction, but be sure to use a trap to prevent water from being drawn into the filtrate flask. Recover the filtrate and discard the particulate material on the filter. Pour the filtrate into a beaker in the hood. Then stir the DMSO–PPO filtrate with a glass rod and very slowly pour about ⅙ vol. of distilled water into the beaker. The result should be a thick white slurry. If a densely packed solid forms, you are pouring too rapidly. Filter the thick white slurry onto Whatman #1 paper as before. This time the PPO will remain on the filter. Wash it with about 5 vols of 10% ethanol made by diluting 95% ethanol into distilled water. Remove most of the liquid from the PPO by aspiration and then spread the compound onto the plastic side of laboratory spill paper. Use a spatula to break up the large lumps and then let it dry for 3–4 days at room temperature.

Chapter 5

Working with Nucleic Acids

This chapter deals almost entirely with DNA methodology. Most of the methods, however, also apply to RNA and will be useful background for the latter half of Chapter 6, which describes a number of procedures that use RNA.

MEASURING NUCLEIC ACID CONCENTRATION AND PURITY

Optical Methods

Both RNA and DNA absorb ultraviolet light so efficiently that optical absorbance can be used as an accurate, rapid, and nondestructive measure of their concentration, even at levels as low as 2.5 μg/ml. A very useful approximation of this absorbance is that double-stranded RNA and DNA at 50 μg/ml in aqueous solution have an $A_{260\,nm}^{1\,cm}$ of 1. This value varies slightly with the (G + C)% of the nucleic acid (Allen et al., 1972), but such variation rarely needs to be considered in molecular biology.

Absorbance is also useful as a measure of the purity of nucleic acid preparations. The absorbance spectrum of double- and single-stranded structures and the increase in absorbance (hyperchromicity) during the transition from double- to single-stranded forms are both fairly accurate measures of purity, but once again are slightly dependent on (G + C)% (Felsenfeld, 1971). In most cases, we use absorbance to estimate the relative purity of different preparations of the same nucleic acid. Thus, for example, if a particular phage lambda DNA preparation was a good substrate in a reaction, then its spectrum and hyperchromicity will be useful criteria for estimating whether another phage lambda DNA preparation is sufficiently clean for another reaction of the same type.

In general, a hyperchromicity of 35% at $A_{260\ nm}$ is expected for pure DNA. The absorbance spectrum, however, is usually the preferred measure of purity because it is more rapidly and easily determined. It also does not require denaturation of the RNA or DNA. The relevant spectrum is between 230 and 320 nm. Absorbance at 320 nm that is more than several percent of the absorbance at 260 nm indicates the presence of undesirable foreign material. In DNA preparations from some sources, for example, plasmid DNA isolated from *E. coli,* such contamination is rare. In preparations from higher organisms such contamination is frequently found and the absorbance at 320 nm can be as high as 10% of the absorbance at 260 nm. If the absorbance at 320 nm is derived entirely from light scattered by particulate matter, a concentration measurement based solely on the absorbance at 260 nm would be in serious error because light scattering increases as the fourth power of the wave number. The ratio of $A_{260}:A_{280}$ of a pure double-stranded DNA preparation should be between 1.65 and 1.85 unless the DNA has a very bizarre (G + C)%. Higher ratios are often due to RNA contamination and lower values to protein or phenol contamination.

Fluorescence Method

The fluorescence of ethidium bromide is increased about 50-fold when it is intercalated into DNA. Since the fluorescence is excited by ultraviolet light and the emission is in the visible spectrum, very simple methods can be used to detect DNA. Further, the high quantum yield provides high sensitivity. In the method described below, for example, 2.5 ng of DNA can be detected and quantitated. In contrast to the UV absorbance, however, this test destroys the DNA.

A spectrofluorometer could be used to quantitate the increased fluorescence of an ethidium bromide solution caused by the presence of DNA. However, we find it considerably easier and sufficiently precise for most practical purposes to make a set of standards and to visually compare them to the unknown sample. The comparisons are made directly by observing the fluorescence with a shielded eye or indirectly with photographs of the sample and standards.

RECIPE

Stock buffer

(this stock buffer remains good for months in the refrigerator)

0.01 M Tris-HCl, pH 7.4
20 mM NaCl

1 mM EDTA
1 μg/ml ethidium bromide

PROCEDURE

Place a number of 20-μl drops of the stock buffer on the surface of a short-wavelength (260 nm) UV transilluminator (Mineralight, Ultraviolet Products, Inc., San Gabriel, Calif.). Add 5 μl of a standard DNA solution (10 μg/ml) to the first drop, mix thoroughly, and transfer 5 μl to the next drop. The same pipette tip can be used to make a total of 5 such serial dilutions if the mixing is done by flushing the sample into and out of the tip several times. Make a similar dilution set of the DNA of unknown concentration. Observe the DNA in the 2 sets after the transilluminator is turned on. Use suitable eye protection when looking at the UV light. Plastic safety glasses are commonly used, but we recommend a large 6-mm thick Plexiglas shield that will protect the eyes and skin. Record the fluorescence of the sets by the photographic procedures detailed in the section on gel electrophoresis (p. 121). Visual interpolation of the fluorescence from the unknown sample into the series containing the standard is reasonably accurate.

COMMENTS

1. Be aware that a wide variety of substances effectively quench the fluorescence. The 2 most common ones are SDS and a contaminant which elutes from some batches of DEAE-cellulose. Whenever there is doubt, add a known quantity of DNA to a drop containing the sample in order to test that the increase in fluorescence is as expected.
2. Drops of the stock solution need not be put directly on a transilluminator. Often it is more convenient to put drops in the top or bottom half of a plastic petri plate and then to invert the plate with a deft twist of the wrist in a way that does not disturb the drops. The inverted plate can then be put on an upward-shining UV source. It is also convenient to stretch Saran Wrap over a ring, for example a truncated tin can (Figure 5.1). Drops can be applied to the Saran Wrap. Despite the high absorbance of UV by plastic, Saran Wrap is so thin that its absorbance is insignificant. Bear in mind, however, that upon extended exposure to UV Saran Wrap, like many plastics, solarizes and fluoresces and, as a result, obscures fluorescence from the sample.
3. The ethidium bromide drop test is not specific for double-stranded DNA. Single-stranded DNA also produces a signal, and can be measured at about half the sensitivity as for double-stranded DNA. RNA also stimulates the fluorescence of ethidium bromide and can be measured at low concentrations.

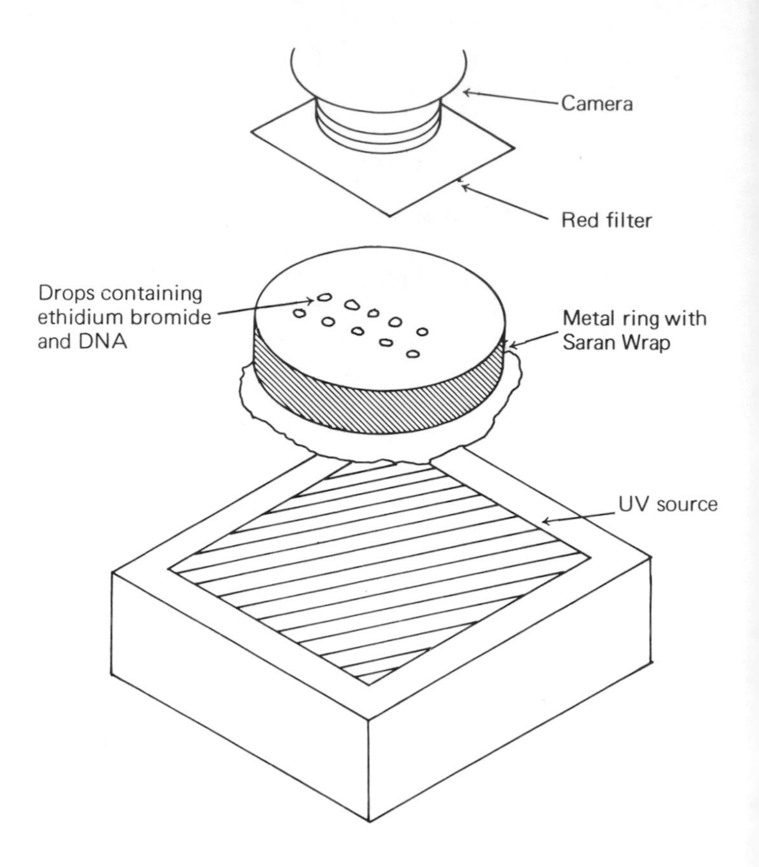

Figure 5.1. An apparatus for quantitating DNA by photographing the fluorescence of ethidium bromide-DNA drops.

4. Ethidium bromide bleaches upon prolonged UV illumination.

STORING DNA

With proper precautions, clean DNA can be stored for long periods of time with little degradation. Storage at 4 °C in 10 mM Tris, pH 8.0, 1 mM EDTA (TE) for 6 months produces less than 1 single-strand break (nick) per 3×10^5 base pairs of supercoiled DNA. A pH of 8 is used because the DNA deamination rate is lower than if the DNA is stored at pH 7. The presence of EDTA eliminates free divalent cations which are required for the activity of most nucleases. The EDTA also inhibits the growth of microorganisms which may synthesize large quantities of nuclease and/or shift the sequence of the DNA to the sequence of the contaminating organism. We have had no problems when using these conditions. Some people have reported trouble with fungal growth in DNA solutions. This can be eliminated by adding a drop of chloroform.

DNA can also be stored with even a lower nicking rate if it is kept frozen in the above solution. However, it appears that every cycle of freezing and thawing introduces some nicks. This can be a serious problem since most −20 °C freezers are equipped with automatic defrost timers. You can disconnect the timer of the defrost unit and all will be well with the DNA, but you will have to do some defrosting from time to time.

CLEANING DNA

DNA preparations are frequently contaminated with substances that inhibit enzymes, degrade DNA, or cause artifactual rates of DNA migration in gels or a variety of other problems. In some cases simple methods can be used to overcome these problems. For example, if the DNA is degraded in less than a day at temperatures between 5 and 40 °C and at a pH between 6 and 9, then nuclease contamination is almost always the problem. This contamination may, in turn, be due to contaminating organisms. Phenol extraction followed by ethanol precipitation is usually the best response to this problem. An alternative response is to add EDTA to 1 mM in order to chelate the divalent cations that are almost always required by nucleases and contaminating organisms. Another common problem, particularly with DNA preparations from eukaryotic cells, is the presence of high molecular weight carbohydrates. These can inhibit enzymatic reactions and usually can be eliminated by pelleting the carbohydrate at 15,000 × *g* for 15 min at 4 °C.

The ultimate recourse is to clean the DNA by column chromatography with DEAE-cellulose (Hirsh and Schleif, 1976) or hydroxyapatite (Martinson, 1973; Britten et al., 1974). In these chromatographic methods the DNA is selectively bound to the column material; contaminants are washed away and then DNA is eluted. The DEAE chromatography is easier and yields cleaner DNA, but had the drawback that longer DNA binds more tightly and is selectively lost. The DEAE method we describe gives about a 40% recovery of DNA 6 kb long and almost 100% recovery of DNA that is 1 kb or less in length. The virtue of hydroxyapatite chromatography is that it works with much larger DNA. Its drawbacks are that it yields slightly less clean DNA and that this DNA is in a highly concentrated phosphate solution. The phosphate cannot be removed by ethanol precipitation of the DNA because the phosphate also precipitates. For this reason removing phosphate is time-consuming. Dialysis is usually the best method but phosphate dialyses very slowly. Occasionally the volume of the solution is small enough that the phosphate can be removed by Sephadex gel filtration.

Note that both of these column methods can be used on highly diluted DNA solutions. Since the DNA is recovered in a small volume, these methods may be used to concentrate DNA.

Cleaning by DEAE

Figure 5.2. A Pasteur pipette prepared for cleaning DNA by passage through DEAE.

PROCEDURE

1. Use washed, preswollen microgranular DEAE-cellulose (DE 52, Whatman) without prewashing. It binds approximately 25 μg of DNA/ml of column volume. Put a substantial wad of siliconized glass wool into the shoulder of a Pasteur pipette (Figure 5.2). Siliconization of the glass wool is only necessary when submicrogram quantities of DNA are to be cleaned. Siliconization reduces the small amount of DNA that will bind to the glass surface of the wool. This bound DNA will not be removed by the treatments we describe.
2. Soak the DEAE in 10 mM Tris-HCl, pH 7.5, 0.3 M NaCl for at least 5 min. Then pipette about 1 ml of slurry into the Pasteur pipette mounted on a ringstand.
3. Wash the column with at least 4 column volumes of the same solution. Allow the solution to flow under the force of gravity.
4. Load the DNA onto the column in a solution that has the same ionic strength.
5. Wash the column 5 times with 2 ml of the same solvent. Elute the DNA with 2.5 ml of 1 M NaCl, 10 mM Tris–HCl, pH 7.4.
6. Dilute this DNA solution with water to 0.2 M NaCl, add 2 vols of ethanol, and precipitate the DNA.

COMMENTS

1. If DEAE-cellulose is eluted with the DNA, it can be removed by a short centrifugation in an Eppendorf microcentrifuge. Next time, pack the glass wool more tightly.
2. We suspect that adherence to the manufacturer's recommendations for the degassing of DE 52, that is, aspiration of the DEAE-cellulose while at low pH in order to remove adsorbed CO_2, leads to substantially reduced DNA recoveries.
3. For the most finicky applications, it is useful to determine the lowest salt concentration which will elute the DNA sample from the column. Often we find the column can be washed with 0.4 M salt and DNA eluted with 0.6 M salt. This procedure should lead to cleaner DNA.

Cleaning by Hydroxyapatite

PROCEDURE

1. Make the phosphate buffer stock (PB), 1.2 M, by mixing 2 vols of 1.2 M Na_2HPO_4 with 1 vol. of 1.2 M NaH_2PO_4 and then adjust to pH 7.0 by adding one or the other of the 1.2 M salt solutions.

2. Hydroxyapatite (HTP, DNA grade, Bio-Rad) is adequate for this procedure after the fine particles (fines) are removed. The fines will clog the column and reduce the flow rate. To remove the fines, suspend the dry hydroxyapatite (HAP) in 0.12 M PB by swirling and then let it settle for approximately 2 min. Remove the supernatant by aspirating with a Pasteur pipette connected to a lab sink aspirator. Repeat this procedure twice and add 0.12 M PB to achieve a total volume that is approximately twice the volume of the settled HAP.
3. Swirl this HAP preparation and add 0.5 ml of the resulting slurry to a water-jacketed (60 °C) column with a diameter of approximately 1 cm and a sintered glass filter at its bottom. Let the HAP sit for 1 min. in the column; then stir and remove the liquid by applying pressure to the top of the column. This can be done with a syringe attached to a needle that sticks through a rubber stopper that, in turn, is pressed into the top of the column.
4. Add the DNA in at least 1 ml of 0.012 M PB, pH 7.0, and let it drain through the HAP with no pressure applied. If the DNA is in a very large volume, stir the mixture. When the draining is complete, apply pressure with a syringe to remove all liquid. On all subsequent washes and elutions, remove all liquid similarly.
5. Elute single-stranded DNA with 2 vols of 2 ml 0.12 M PB, 60 °C, or elute both single-stranded and double-stranded DNA with 2 vols of 2 ml 0.4 M PB, 60 °C.
6. Combine the 2 vols of 2 ml and remove the phosphate by dialysis against a volume of 1 liter. Dialyse twice against 0.2 M NaCl, TE, pH 8 for 2 h and then twice for at least 1 day against the same solution. Ethanol precipitate the DNA. The 0.2 M NaCl accelerates the dialysis rate of the highly charged phosphate, presumably because it shields the charge interactions between the phosphate and dialysis membrane.

COMMENTS

1. The binding capacity of DNA-grade HAP is approximately 0.6 mg of DNA/ml of bed volume. However, this capacity does vary with different commercial lots.
2. If water-jacketed columns are not available, the salt concentration can be adjusted so that this fractionation can be done at room temperature with non-DNA-grade Biogel HTP (Martinson, 1973).

PRECIPITATING DNA WITH ETHANOL

Ethanol precipitation is the standard method for concentrating DNA or changing its solvent. Be aware that EDTA above several millimolar will precipitate and hence can be carried into the next reaction. Since phosphate precipitates at low concentrations, it is desirable to design experiments avoiding phosphate.

This method is good for precipitating nanogram quantities of DNA provided that all glassware is siliconized or plastic tubes are used.

PROCEDURE

1. Add 1 M Mg^{2+}-acetate to make the solution 10 mM and add 3 M Na^{+}-acetate to make the solution 0.3 M. Then add 2.5 vols of 95% ethanol. Mix thoroughly and then freeze in a dry ice–ethanol bath for 10 min if the DNA is in plastic 1.5-ml microfuge tubes or for 20 min if it is in glass 15- to 50-ml Sorvall centrifuge tubes. Do not be concerned if the liquid freezes or forms a thick slurry. Spin at 10,000 rpm for 15 min in the Sorvall SS-34 rotor or for 10 min in an Eppendorf microcentrifuge. Remove the supernatant.
2. If highly radioactive triphosphates or any similar moderate-sized molecule is to be removed from the DNA, resuspend the pellet in 0.3 M Na^{+}-acetate, pH 5.5, add 2.5 vols ethanol, mix and freeze, and spin as above.
3. Wash the pellet by adding 2 ml 70% ethanol and then inverting the tube several times. Freeze, spin 5 min, remove supernatant, and dry in a vacuum desiccator. Complete drying of the pellet greatly facilitates resuspending the DNA in small volumes of buffer but is not necessary when resuspending in large volumes.

COMMENTS

1. In many instances larger quantities of DNA ($> 0.1\ \mu g$) are efficiently and more conveniently precipitated by bringing the NaCl concentration to 0.25 M, adding 2 vols of 95% ethanol, either chilling in dry ice–ethanol for 15 min or storing overnight at -20 °C, and then pelleting the precipitate by 30-min ultracentrifugation of large volumes at 100,000–200,000 $\times$ *g* or centrifugation of small volumes in the Eppendorf microfuge.
2. Very large quantities of DNA ($\geq 50\ \mu g$) can be precipitated at lower *g* force in the Sorvall preparative centrifuge.
3. Occasionally subnanogram quantities must be precipitated. This is most efficiently done by adding carrier tRNA to about 50 μg/ml before using the above procedure.
4. DNA larger than 10 kb resuspends very slowly. It may take a day and requires gentle agitation.

TCA PRECIPITATION ASSAY

Precipitation with trichloroacetic acid (TCA) is widely used to separate radioactive oligomeric nucleic acids from radioactive nucleosides and nucleotides. This is accomplished by a simple procedure which precipitates the oligomers and collects them on a filter. Oligonucleotides of length greater than 10–20 nucleotides are retained on the filter. If the exact minimum size precipitated is important, it should be checked.

PROCEDURE

1. Add carrier tRNA, calf thymus DNA, or bovine serum albumin to the sample if it contains less than about 20 μg/ml of nucleic acid. Any quantity of carrier between 20 and 500 μg/ml seems to work satisfactorily. If the radiolabel is ^{3}H, use 20 μg/ml of carrier to reduce quenching.
2. Add 2 ml of 5% TCA at 0–4 °C to 0.5 ml or less of the nucleic acid solution. Vortex and leave on ice for 10 min.
3. Filter through a GF/C glass filter (Whatman) and then rinse the filter with 5 ml of 1% TCA, 0–4 °C. When there is a large excess (≥ 100-fold) of labelled nucleosides compared to labelled nucleic acid, background due to nucleosides binding to the filter can be a problem. The background can be substantially reduced by including 0.1 M pyrophosphate in the 5% TCA solution and by soaking the GF/C filters in this 5% TCA, 0.1 M pyrophosphate solution for at least 5 min before precipitating onto the filters.
4. Rinse with 2 ml of 95% ethanol, dry, and count.

PRECIPITATION AND SIZE FRACTIONATION OF DNA WITH POLYETHYLENE GLYCOL

Polyethylene glycol (PEG; Carbowax 5000, Union Carbide) precipitates DNA in a nearly quantitative manner provided that the initial DNA concentration is greater than 50 μg/ml. The method is inexpensive, has virtually unlimited capacity, does not harm the DNA, and is easy to perform. Precipitation by PEG has the very useful additional feature that it can be manipulated to selectively precipitate different sizes of DNA molecules utilizing different concentrations of PEG. For example, two DNA fractions that differ in size by a factor of 2 and that are within the length range of 100–2000 base pairs can be separated with only about 5% cross-contamination. This is useful for bulk purification before quantitative purification by electrophoretic techniques. The same method can also be used to separate plasmid DNA from chromosomal DNA. Most of the relevant experimental variables for precipitation by PEG have been investigated (Lis and Schleif, 1975), so that it is possible to predict rather closely the conditions necessary to achieve the desired fractionation. However, it is advisable to perform pilot experiments in order to maximize the separation for any particular application.

We describe below a procedure for separating 2 EcoRI restriction fragments that are approximately 440 and 5000 base pairs, respectively, in length. The 4-base cohesive ends on these fragments present a problem not addressed in the publication mentioned above. Whenever possible, DNA manipulations are performed at 0 °C in order to minimize unwanted chemical and enzymatic degradation. However, the cohesive ends remaining after EcoRI endonuclease digestion allow the fragments to aggregate at 0 °C. Therefore size fractionation does not work at this temperature. The problem is overcome by performing the PEG precipitation at 37 °C. This prevents the cohesive ends from annealing and hence prevents aggregation.

PROCEDURE

1. Heat the sample to 65 °C for 10 min to inactivate the EcoRI enzyme.
2. Dilute the sample 10-fold with TE buffer in order to reduce the concentration of Mg^{2+} to below 0.5 mM. The resulting solution must contain, however, at least 50 μg DNA/ml.
3. Add 5 M NaCl to give a final concentration of 0.55 M.
4. Add and dissolve 6 g PEG per 100 ml.
5. Incubate at 37 °C for 20 h.
6. Centrifuge at 10,000 rpm at room temperature in a Sorvall GSA rotor for 10 min. The supernatant contains the 440-base pair fragment. The pellet contains the 5000-base pair fragment.
7. To precipitate the 440-base pair fragment from the supernatant, add and dissolve PEG to make a final concentration of 18%.
8. Incubate at 0 °C for at least 2 h.
9. Centrifuge at 10,000 rpm at 0–4 °C in a Sorvall GSA rotor.
10. Resuspend pellet in TE buffer.
11. Precipitate the DNA in ethanol and resuspend in TE or another desired buffer.

ISOLATING *E. COLI* DNA

The isolation of DNA from *E. coli* poses relatively few problems. With care it is possible to isolate very pure DNA with an average fragment size above 10 kb. Below is given a modification of the Sato and Miura (1963) method, which yields DNA adequate for cloning by recombinant DNA technology or for other procedures that require enzymatic operations on the DNA.

PROCEDURE

Unless otherwise noted, all operations are done at 0 °C with solutions that have been chilled to 0 °C.

1. Grow cells overnight to stationary phase in 5 ml of YT medium.
2. Centrifuge 3 × 1.5 ml of cells in the Eppendorf centrifuge.
3. Resuspend the combined cell pellets from the 4.5 ml of cell culture in 15 μl of lysozyme solution (freshly dissolved at a concentration of 2 mg/ml in 0.15 M NaCl, 0.1 M EDTA, pH 8.0.
4. Incubate at 37 °C for 10–20 min until the cells just begin to lyse. Lysis is indicated by an increase in viscosity.
5. Freeze in dry ice–ethanol.
6. To the frozen cells, add 125 μl of 1% SDS, 0.1 M NaCl, 0.1 M Tris-HCl, pH 9.0. After adding, stir as the cells thaw.
7. Add 150 μl of phenol that has been saturated with the buffer described in step 6 and then mix.
8. Centrifuge 3 min in an Eppendorf centrifuge.
9. Remove the upper aqueous phase which contains the DNA, and add 300 μl of 95% ethanol. Mix, freeze at −20 or −70 °C for 30 min, and centrifuge in the Eppendorf centrifuge for 5 min.
10. Resuspend the DNA pellet in 100 μl of 0.1× SSC buffer (see Appendix I, Commonly Used Recipes). The low concentration of salt promotes dissolving of the DNA. After the DNA has dissolved add 5 μl of 20× SSC buffer.
11. Add 4 μl RNase TI (Sigma, R8521) at 0.8 mg/ml and 3 μl RNase A (Sigma, R4875) at 2 mg/ml and incubate at 37 °C for 30 min.
12. Add 100 μl of phenol saturated with SSC, vortex, and centrifuge 3 min.
13. Remove the aqueous layer and precipitate the DNA in ethanol.

ISOLATING LAMBDA DNA

Two basic methods exist for the isolation of DNA from purified lambda phage particles. One removes protein by phenol extraction and the other removes it by sodium dodecyl sulfate (SDS) precipitation. Rumors about the impossibility of removing the last traces of SDS from the DNA have led us to use the SDS method only when the DNA need not be scrupulously clean. We use phenol extraction when the DNA will be used for genetic engineering, that is, when it is to be cut with restriction enzymes, ligated, kinased, and so on.

Phenol Extraction

This is a phase-partition extraction in which the nucleic acids partition into the aqueous phase and proteins, mostly denatured, partition into the more dense phenol phase and into the interface (Kirby, 1957).

PROCEDURE

1. Dialyse the phage against lambda DNA buffer (0.01 M Tris-HCl, pH 8.0, 0.1 M KCl, 10^{-4} M EDTA) for 1 h at 4 °C in #20 dialysis tubing. During the dialysis many phage will pop and the solution will become more viscous.
2. Before or after dialysis determine the concentration of phage DNA. Note that the approximate absorbance of DNA (50 μg/ml has an OD_{260}^{1cm} of 1) applies whether or not the DNA is in the phage coat.
3. Use lambda DNA buffer to adjust the DNA concentration to 200 μg/ml. Extractions at higher concentrations are possible, but they are more awkward due to the increased viscosity of the solutions.
4. Neutralize distilled phenol by shaking it with 2 changes of 0.2 vol. of 1 M Tris-HCl, pH 8, and 2 changes of DNA buffer.
5. Extract the DNA by adding an equal volume of the neutralized phenol to the DNA solution and then gently mixing or shaking at 4 °C for approximately 20 min. This length of time is necessary when several ml are being extracted. If a volume as small as 50 μl is being extracted, 1 min of mixing is enough.
6. Separate the phases by centrifugation at 5000 × g for 1 or 2 min and then by gently removing the upper phase. This is most easily done with a bent-tipped Pasteur pipette (Figure 5.3) in order to keep from drawing up the material at the interface between the phenol and water.
7. Reextract with another equal volume of neutralized phenol.
8. Dialyse against lambda DNA buffer to remove the phenol. Three changes of buffer with 8 h dialysis each time is adequate. Residual phenol can be detected spectrophotometrically because a solution saturated with phenol has an $OD_{270}^{1\,cm}$ of about 1100. However, for most people, the nose is a more sensitive assay for residual phenol. Phenol dialyses slowly and requires these long dialyses. The removal of phenol can be accelerated by extracting the DNA solution 2 to 4 times with chloroform immediately after phenol extraction. The phenol partitions into the chloroform phase. The residual chloroform can be removed by brief dialysis or by an ethanol precipitation step.

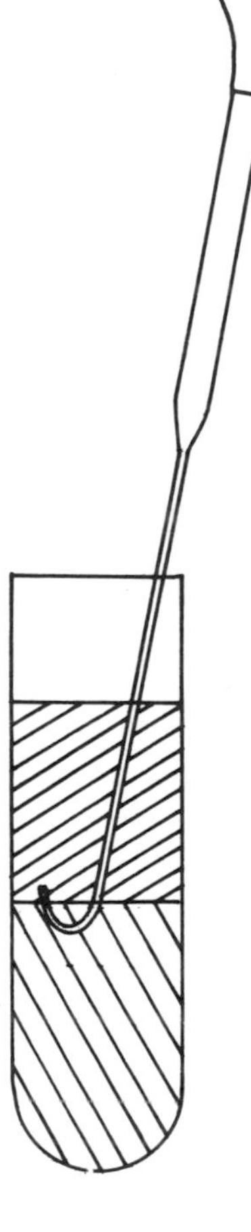

Figure 5.3. The use of a bent-tipped Pasteur pipette to remove the viscous DNA phase from the protein interface and phenol lower phase during phenol extraction.

SDS Extraction

This extraction utilizes SDS to denature coat protein and release the phage DNA. The protein and most of the SDS are then removed by precipitating the SDS with K^+ because a potassium-SDS complex is highly insoluble. In general, K^+ must be avoided in all solutions containing SDS unless SDS precipitation is the objective.

PROCEDURE

1. Dialyse phage into 0.05 M NaCl, 0.01 M Tris-HCl, pH 7.5-8.0, 0.001 M Na-EDTA.
2. Add SDS (reasonably pure or recrystallized from ethanol) to make 0.5%.
3. Incubate at 65 °C for 10-15 min. During this incubation the solution should clear and become viscous.
4. Add KCl to make a concentration of 0.5 M and chill at 0 °C for 15 min.
5. Centrifuge out the K^+-SDS in the cold. The required *g* force and the time of centrifugation are dependent upon the viscosity of the solution, but a good starting point is 7000 × *g* for 10 min. The Sorvall SS34 rotor is satisfactory.
6. Dialyse the supernatant at 4 °C into 0.1 M NaCl, 0.01 M Tris-HCl, pH 7.5-8.0, 0.001 M EDTA. Dialyse for at least 12 h with several buffer changes using #20 dialysis tubing. After this time the buffer can be changed to one containing KCl instead of NaCl.

ISOLATING PLASMID DNA

This procedure is very reliable and always yields DNA that can be efficiently acted upon by even the most finicky of enzymes. The procedure is essentially that of Clewell and Helinski (1969) and has 4 steps. The first increases (amplifies) the number of plasmid copies per cell. The second purifies the plasmid from most bacterial material. The third separates supercoiled plasmid DNA from other DNA and from most of the remaining bacterial debris. The final step removes contaminating low molecular weight material by a simple gel filtration procedure.

Plasmids can be divided into 2 classes according to their mode of replication. The most useful class for genetic engineering manipulations has relaxed control of DNA replication. Plasmids of this class, unlike the bacterial host DNA, do not require continued protein synthesis for their replication. For this reason, in the first step of this isolation procedure these plasmids may be selectively amplified simply by adding a protein synthesis inhibitor to the bacterial culture (Clewell, 1972). In practical terms this leads to a 10- to 20-fold increase in the yield of plasmid.

The second step of this procedure gently lyses the bacteria with a combination of mild detergents. A fairly hard centrifugation forms a precipitate which includes most of the chromosomal DNA of the bacterial host as well as most of the membranes and other large structures. The plasmid DNA and other small molecules are found in the supernatant.

In the third step, ethidium bromide and CsCl are added to the supernatant and the resulting solution is centrifuged to equilibrium in an ultracentrifuge (Radloff et al., 1967). Ethidium bromide intercalates between the stacked bases of the DNA helix and lowers the density of the helix. Covalently closed supercoils cannot intercalate as much ethidium bromide as can linear and open circular DNA. As a result, the covalently closed molecules of the plasmid form a distinct higher density band in a density gradient. Nicked circles and linear chromosomal DNA form a lower density band. The same density gradient centrifugation step pellets large RNA molecules and raises proteins, lipids, and many other molecules to the top of the gradient, forming a pellicle. Small molecules of RNA are found throughout the gradient and can account for a substantial fraction of the A_{260} material in the plasmid DNA band.

If a relaxed replication plasmid is used and is amplified, the plasmid yield can be 0.5–1.0 μg of DNA/ml of bacterial culture. However, the yield can vary substantially from one plasmid to another as well as from host to host.

Plasmid-containing strains have a tendency to lose their plasmids. For this reason an easily selectable marker is usually included in most plasmids used in the laboratory. Some trouble will be avoided by selecting for such a marker during growth of these strains.

In the procedure described below we assume that the plasmid to be isolated has relaxed control of replication and codes for tetracycline resistance.

PROCEDURE

1. Grow an overnight culture in 5 ml of LB with 25 μg/ml tetracycline.
2. In the morning use the entire overnight culture to inoculate 600 ml of the same medium with the same antibiotic in a 2-liter flask. Grow cells to approximately 4×10^8 cells/ml (about 3 h) and add chloramphenicol, solid or dissolved in H_2O, to yield a final concentration of 250 μg/ml. Incubate the cells with shaking at 37 °C for more than 22 h. Shorter chloramphenicol treatment will lead to substantially lower yields and treatments as long as 36 h give no substantial change in the yield.
3. If desired, or if necessary to follow recombinant DNA guidelines, add 1.5 ml $CHCl_3$ to the culture and shake for 5 min to kill any possible survivors. This treatment kills cells in the culture liquid and in the aerosol within the flask.
4. Pellet the cells by centrifugation at $5000 \times g$ for 10 min, 4 °C. Discard the supernatant.
5. Resuspend cells in 100 ml of TE, pH 8, 4 °C. Throughout this recipe the pH measurements are to be made at room temperature. The temperature given refers to the temperature at which the buffer is used. Repellet, decant the supernatant as before, and resuspend the pellet in 2 ml of 25% sucrose, 50 mM Tris–HCl, pH 8, 4 °C. Put the resuspended pellet in an ice water bath and

keep it in the bath through step 7. Add 0.4 ml of 10 mg/ml lysozyme in 0.25 M Tris–HCl, pH 8, 4 °C. Mix the cell suspension thoroughly but gently with a glass rod and leave the suspension in the ice bath for 10 min. The lysozyme digests the peptidoglycan layer of the cell wall.

6. Add 1 ml of 0.25 M EDTA, pH 8, 4 °C. Mix the resulting solution with a glass rod as in the previous step and leave in the ice bath for another 10 min. The EDTA will chelate divalent cations and prevent nuclease degradation of DNA in subsequent steps. The chelation also weakens the outer membrane of the cell wall.
7. Add 3.2 ml of 1% (wt/vol) Brij 35 (Calbiochem), 0.4% (wt/vol) deoxycholate (Calbiochem), 0.06 M EDTA, 50 mM Tris–HCl, pH 8, 4 °C. The solution will become viscous at this step. Leave it in the ice bath for 10 min and stir thoroughly but gently with a glass rod for the entire 10 min. This thorough mixing appears crucial to high plasmid yields. Presumably it allows the detergents to reach most of the cells and not be trapped in viscous regions of the liquid.
8. Centrifuge at 48,000 × *g* for 30 min, 4 °C. Steps 5–8 are conveniently done in a 50-ml polypropylene screw cap (Oak Ridge style) tube (Sorvall) with centrifugations in the SS34 rotor of a Sorvall preparative centrifuge. With this rotor the centrifugation would be at 20,000 rpm.
9. Decant the supernatant into a 50-ml flask that contains 6.8 g of CsCl. Warm the solution to room temperature and when the CsCl is in solution add 0.5 ml of 15 mg/ml ethidium bromide. Adjust the refractive index at 25 °C to between 1.3930 and 1.3960 by adding distilled water or solid CsCl. With the volumes and quantities used in this recipe, a decrease in refractive index of about 0.006 is produced by the addition of 1 ml H_2O. Use plastic or rubber gloves for this and subsequent operations, as ethidium bromide is probably a carcinogen. After the refractive index has been adjusted, let the solution sit for a few minutes, mix it, and check that the refractive index has not changed. Solid CsCl can take a while to dissolve. An alternative and simpler method for obtaining the proper density is to bring the volume of the supernatant to 9 ml by adding water and then add 9 g of CsCl and 0.4 ml of the ethidium bromide solution. If these proportions are used, no refractive index measurement need be made.
10. Centrifuge at 130,000 × *g*, 20 °C, for at least 20 h in a fixed angle rotor. The volumes described above allow all of the plasmid DNA preparation from a 600-ml culture to be centrifuged in a single tube of a Beckman TI 50 rotor. With this rotor centrifugation would be at 45,000 rpm. Centrifugation for up to 40 h is necessary to obtain sharp DNA bands if the DNA yield is high and therefore the viscosity is also high. It can also be due to

minor variations in steps 7, 8, or 9 that lead to variations in the amount of contaminants with high affinity for DNA. In our experience vertical tube ultracentrifuge rotors will give adequate separation of plasmid and host DNA after 4–8 h of centrifugation at the maximum rotor speed for a solution of this density.

11. After centrifugation, clamp the tube to a ringstand and illuminate it with a long-wave UV spotlight (Blak-Ray B100A, Ultra Violet Products, Inc., San Gabriel, California). Open the top of the tube to atmospheric pressure and remove the lower fluorescing band with an 18-gauge needle attached to a 1-ml syringe. Puncture the tube about 5 mm below the lower band and lever the needle up until its tip, with the beveled side up, is just below the lower band (Figure 5.4). Remove the band withdrawing $\leqslant$ 1 ml from a 10- to 12-ml gradient. The band will never fully disappear because as the solution is withdrawn a discontinuity is created in the CsCl gradient which refracts with a pinkish glow that looks like a DNA band. Use gloves for this because of the ethidium bromide. In rare instances the gradient has large particles in the DNA region. If a gradient has these, centrifuge the DNA solution at 15,000 $\times$ g for 15 min after step 13 and discard the pellet. Save the CsCl from the rest of the gradient, and recycle it (p. 188) for future gradients.

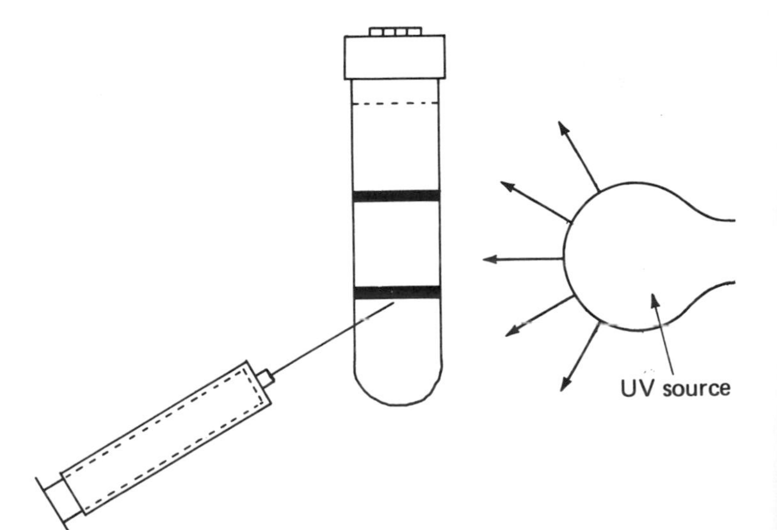

Figure 5.4. Collecting the plasmid DNA band following equilibrium gradient centrifugation in cesium chloride-ethidium bromide.

12. Extract the ethidium bromide with an equal volume of water-saturated *n*-butanol at room temperature in a chemical hood. Approximately 6 extractions are necessary. Follow the extraction with the long-wave UV spotlight and extract an additional 2 times after the ethidium bromide is no longer detectable. The aqueous volume will increase during these extractions. If the volume is

too large for subsequent procedures it can be reduced to the original 1 ml by extracting with *n*-butanol that is not water-saturated. Unsaturated *n*-butanol is not recommended for all extractions because its use will increase the concentration of CsCl to the point that CsCl will precipitate.

13. The following step will remove CsCl, low molecular weight RNA, and other low molecular weight contaminants. This is necessary since the quantity of low molecular weight RNA contamination is unpredictable and can be greater than the mass of plasmid DNA. Layer not more than 1.5 ml of extracted DNA onto a 7.5-ml Bio-Gel A-15m (100–200 mesh; Bio-Rad) column formed in a Bio-Rad 10-in. disposable column. A new A-15m column should be washed with 40 ml of TE, pH 8, before use; a previously used column should be washed with 20 ml. Collect 20-drop fractions (approximately 0.5 ml) and the excluded DNA will be in fractions 4–9, as shown in Figure 5.5. Assay the fractions spectrophotometrically at A_{260} or, if desired, by gel electrophoresis.

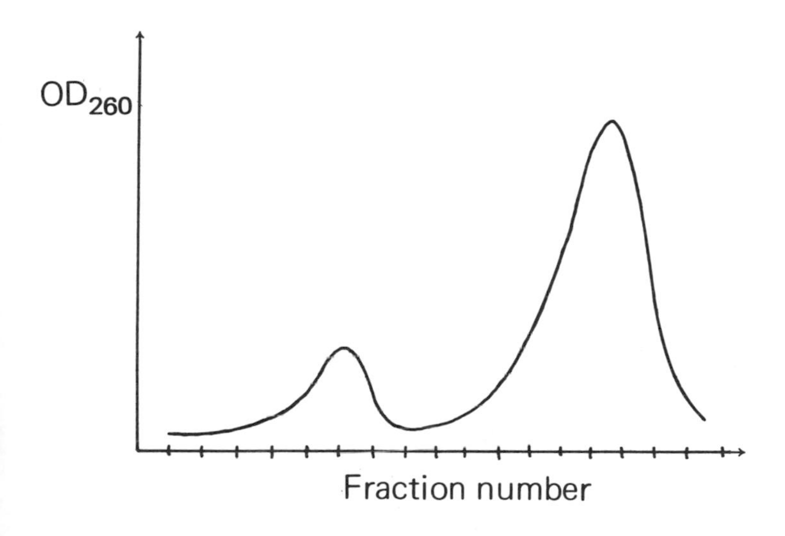

Figure 5.5. A typical elution profile of DNA and small oligonucleotides from an agarose column during plasmid purification.

14. It is usually convenient to concentrate this plasmid DNA about 10-fold by ethanol precipitation. If this is done, small volumes containing sufficient DNA can then be added to enzymatic reactions without appreciably affecting the reaction buffer composition.

COMMENTS

1. The amount of chloramphenicol called for in the recipe can only be described as a whopping amount. As little as 25 μg/ml is satisfactory. Some lots of chloramphenicol give adequate plasmid yields only when used at this lower concentration. Presumably they contain impurities which inhibit DNA replication if the chloramphenicol has been added at the higher concentration.

2. A potential problem is the lysozyme step. Occasionally a lot of lysozyme is rather inactive or does not work well on some days. At the end of the 10-min incubation of step 5 it pays to observe the cells in a microscope. If the treatment has been adequate, at least half, and preferably more, of the cells will be visibly distorted or complete spheroplasts. If too few cells have been converted to spheroplasts, increase the time of incubation and/or the amount of lysozyme.
3. Pure CsCl is not inexpensive and is not required. Chapter 7 describes a procedure that we routinely use to recycle CsCl from preparative density gradients. Occasionally it is possible to purchase inexpensive CsCl from inorganic chemical suppliers. Often this is a great buy, but the amount of contaminating rubidium varies. Consequently the relationship between the refractive index and the position of the banded DNA in the centrifuge tube will vary and you will have to determine your own refractive index for the preparation. We have heard that some batches of inexpensive CsCl contain massive quantities of acid that will degrade DNA. Check the pH of a dissolved sample of CsCl before using it with DNA.
4. It is tempting to postpone the extraction of ethidium bromide until after running the agarose column. This would appear to have the advantage of using fluorescence to display the exact location of the DNA as it comes off the column. For reasons which are unclear it is very difficult, if not impossible, to remove all the ethidium bromide from DNA if this has been done.
5. Ethidium analogs are more expensive but some of them give considerably better separation between supercoiled and linear DNA bands in gradients (Hudson et al., 1977).

LARGE-SCALE PLASMID ISOLATION

At present the best method for obtaining up to 1 mg of highly pure plasmid DNA is the ethidium bromide–CsCl procedure just described. That procedure, however, cannot be easily scaled up to yield substantially larger quantities of plasmid. An alternative method described herein is readily adaptable to any quantity of DNA because it uses precipitation steps to purify the plasmid DNA. This alternative has its drawbacks, however. First, it often yields plasmid DNA contaminated with about 10% chromosomal DNA. This contamination rarely is a problem if additional purification steps are planned, as in, for example, the isolation of a particular restriction fragment from the plasmid. The second drawback is that the resulting DNA is contaminated by large quantities of tRNA. When this contamination is harmful it can be removed by Bio-Gel A-15m gel filtration as described in this procedure or by RNase digestion followed by PEG precipitation of the plasmid DNA.

PROCEDURE

The cells should be grown and the plasmid amplified as described in the previous section. Since the purification procedure described below requires long incubations it is best to prevent the proliferation of extraneous nucleases produced by microbial contaminants. This can be done by sterilizing all solutions either by autoclaving or filtration. The following recipe is described for 10 g (wet weight) of cells.

1. Resuspend the cells in 100 ml of 25% sucrose, 0.05 M Tris-HCl, pH 8.0, at 4 °C. Stir in an ice bath for 30 min or until all cells are suspended.
2. Add 20 ml of 20 mg/ml lysozyme in 0.25 M Tris-HCl, pH 8.0, at 4 °C. Dissolve the lysozyme just before use. After adding the lysozyme, stir the cells with a glass rod for 15 min while keeping them cold in an ice water bath.
3. Examine aliquots of cells in a microscope to determine that most cells have become spheroplasts. If they have not, continue stirring until this has occurred.
4. Add 40 ml 0.25 M EDTA, pH 8.0, at 0 °C and mix gently with a glass rod for 10 min while keeping cells cold in the ice bath.
5. Add 160 ml of lysis mix at 0 °C (see below). This will cause the solution to become quite viscous. Mix and let stand for 10 min on ice.

Lysis Mix

2 ml 10% Triton X-100
50 ml 0.25 M EDTA, pH 8.0
10 ml 1 M Tris-HCl, pH 8.0
138 ml H_2O

6. Pour into autoclaved or ethanol-washed centrifuge bottles and then centrifuge for 2 h at 12,000 rpm in the Sorvall GSA rotor, at 4 °C.
7. After centrifugation there will be 3 phases: a clear supernatant which pours off easily, a rather clear but viscous supernatant that blends into the top supernatant, and an extremely viscous pellet. The top supernatant contains the plasmid. The more viscous supernatant contains much chromosomal DNA and is best avoided when collecting the top phase. Occasionally, yields of plasmid can be increased by recentrifuging the two lower phases and recovering the new top phase.
8. Heat-precipitate some of the proteins by bringing the supernatant to 65 °C for 15 min. Then remove the precipitated protein by centrifuging in the Sorvall GSA rotor for 10 min at 5000 rpm and 4 °C.
9. Begin the DNA precipitation by adding 3.5 ml of 5 M NaCl and 7.0 ml of 50% PEG 6000 (Carbowax 6000) for every 25 ml of supernatant. Mix and let stand on ice for 1 h. Centrifuge in the Sorvall GSA rotor for 20 min at 8000 rpm and 4 °C.

10. Resuspend the pelleted DNA in 20 ml of 10 mM Tris-HCl, pH 7.4, 1 mM EDTA, and 0.2 M NaCl.
11. Ethanol-precipitate the DNA by adding 3 vols of ethanol and chilling to −20 or −70 °C. Although only an hour of chilling is sufficient, this is a convenient stopping point for this preparation. The following centrifugation can be completed the next day. Centrifuge 30 min at 19,000 rpm in the Sorvall SS34 rotor. Pour off the supernatant and resuspend the pelleted DNA in TE buffer.
12. To kill any residual nucleases, cross-link proteins by treating with diethyl pyrocarbonate (DEPC; Sigma D5758) (Berger, 1975). Add DEPC to 0.1%, mix immediately, and incubate for 10 min on ice. Inactivate the DEPC by heating to 65 °C for 10 min. Remove cross-linked proteins by centrifugation for 10 min at 8000 rpm in the Sorvall SS34.
13. Ethanol-precipitate the DNA and resuspend it in the buffer of your choice.
14. At this point in the procedure, tRNA usually constitutes two-thirds of the nucleic acid. The tRNA can be removed by Bio-Gel A-15m gel filtration. Five to 10 mg of plasmid DNA and perhaps even more can be purified on a column that has a bed volume of about 300 ml. To determine the approximate concentration of DNA at this stage, it is most convenient to electrophorese an aliquot of it on a 0.7% agarose gel together with parallel lanes that contain known amounts of DNA. After determining the DNA concentration, make your sample to 0.5 M NaCl, 0.01 M Tris-HCl, pH 7.5, and 10^{-4} M EDTA, and heat to 65 °C for 10 min just before loading up to 1.5 ml of the solution on the column. Preflush the column in the same buffer that has NaCl at 0.2 M rather than at 0.5 M. The column flow rate should be about 1 ml/min and 2-ml fractions should be collected. Measure the elution profile either with the ethidium bromide drop test or by absorbance at 260 nm. It is best to electrophorese the column fractions on an agarose gel before pooling fractions. The gel allows the purity to be determined.

ISOLATING *DROSOPHILA* DNA

A single method can be used to isolate reasonably clean, high molecular weight DNA from the nuclei of almost any higher organism. In this method the nuclei are lysed in a detergent, the protein is digested with protease, the DNA is banded in a CsCl density gradient, and then as an optional step the remaining protein can be removed by phenol extraction. The CsCl banding step seems to be particularly important when the DNA must be clean enough for restriction enzyme digestions and for some other finicky DNA manipulations. However, this banding step does have one potential drawback because unless special care is taken, it is possible to selectively lose part or all of the DNA fractions that have unusual density, as for example the mitochondrial DNA from some species and the simple-sequence satellite DNAs.

If this method of purification is used, the crucial step for obtaining high molecular weight DNA appears to be the isolation of nuclei. This may be due to nuclear membranes becoming permeable to cytoplasmic nucleases and/or breakdown of compartments within the nucleus allowing nuclear enzymes to digest the DNA. The methods for preparing nuclei vary from one species to the next. We describe a method for isolating nuclei from *Drosophila melanogaster* embryos (Gergen et al., 1979).

With this method the highest molecular weight DNA will be isolated when the temperature is kept very close to 0 °C between the time of embryo breakage and of nuclear lysis.

PROCEDURE

1. This procedure is for 50 wet-weight g of 0- to 20-h-old embryos. It has been used for larger quantities of embryos, but high yields are obtained only when all solution volumes are scaled up in proportion to the increase in egg weight. Lower molecular weight DNA ($\leq$ 40 kb) can be extracted from frozen embryos if they have been rinsed in 0.7% NaCl, drained dry, and placed in a −80 °C freezer for freezing and storing. They should be thawed at 37 °C until they become soft.
2. Do the following operations at room temperature. Put the embryos on Nitex 63 cloth (Tetko, Elmsford, N.Y.) that is held between two Plexiglas rings, as shown in Figure 5.6. Wash the embryos extensively with distilled water. Use about 4 liters of water and stir the embryos while pouring the water onto the embryo mass. Wash the embryos with 1 liter of 0.7% NaCl, 0.01% Triton X-100, and use the same stirring procedure. Dechorionate the embryos by placing the rings in a petri dish and pouring a 4-fold-diluted Chlorox solution (Chlorox is 5.25% sodium hypochlorite) into the upper ring. The petri dish forms an air lock and keeps the embryos submerged in the diluted Chlorox. Stir the embryos for 2 min and then drain and wash with 1 liter of the 0.7% NaCl, 0.01% Triton

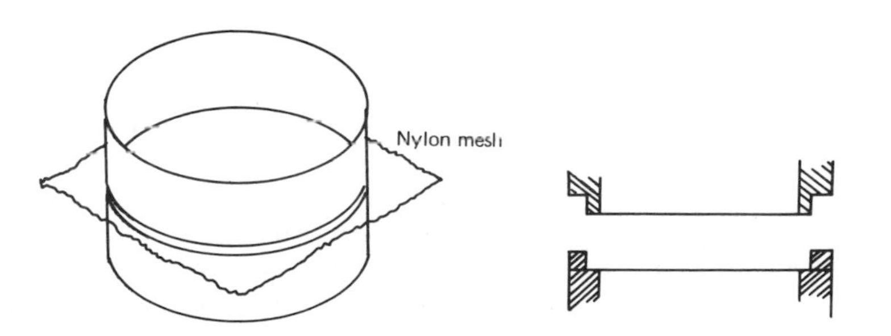

Figure 5.6. A Plexiglas clamp for Nitex cloth which permits washing of *Drosophila* eggs. The apparatus is made oval rather than circular so that the upper and bottom portions may be tightly attached to one another by slightly twisting one with respect to the other.

X-100 solution. Wash the embryos with 100 ml of embryo buffer and then remove most of the moisture by laying the bottom surface of the Nitex on paper towels.

3. Do the following operations in a cold room with all containers sitting on ice. Transfer the embryos to a prechilled mortar and pestle. Grind vigorously with the pestle. After a few minutes of grinding take a small sample from the mortar and examine it in a dissecting microscope to determine the extent of embryo breakage. The embryonic egg shells are white when they are not broken and transparent when they are broken. When more than 90% of the egg shells are broken, add 100 ml of the embryo buffer to the mortar and suspend all of the debris. Pour this resuspension through Nitex 63 in a funnel that drains into a 250-ml Sorvall centrifuge bottle, which in turn is held by ice in a bucket. Squeeze the liquid out of the debris by wrapping the Nitex around it and squeezing with your gloved hands. Rinse the mortar and pestle with another 100 ml of embryo buffer and pour this through the Nitex and squeeze again. In all of these maneuvers it is best to operate quickly. The sooner after egg shell breakage that the nuclei are separated from cytoplasm the more likely you are to obtain high molecular weight DNA.
4. Centrifuge the filtrate in a GSA Sorvall rotor at 13,000 rpm, and 2 °C, for 15 min. Resuspend the pellet in 10 ml of ice-cold embryo buffer. Add 135 ml of ice-cold 50% (wt/wt) sucrose in embryo buffer. Mix this gently but thoroughly and centrifuge at 13,000 rpm, 2 °C, for 1 h in the Sorvall HB4 rotor. Resuspend each nuclear pellet in 4 ml of cold embryo buffer, pool the resuspended pellets, and add 6 ml of cold lysis buffer. Add 1.6 ml of 5 mg/ml pronase and incubate at 50 °C for 1 h. Proteinase K can be substituted for pronase. Proteinase K has very little if any nuclease activity and thus need not be self-digested. It is more expensive, and for most DNA preparations self-digested (see Recipes below) pronase seems adequate.
5. The DNA can be banded in CsCl by either of the following two methods. We find banding in a CsCl–ethidium bromide gradient more convenient because it allows the DNA to be collected through the side of the centrifuge tube using a syringe, as described for plasmid DNA isolation (p. 104). For this method add 22 g of CsCl and 1.6 ml of 15 mg/ml ethidium bromide to the digested lysate. Adjust the refractive index to 1.403 at 20–25 °C and then centrifuge at 42,000 rpm for 40 h in a Beckman TI50 angle rotor. Remove the DNA band by side puncture and extract the ethidium bromide as described for plasmid DNA preparations (p. 104). Dialyse the DNA against TE, pH 8. The alternative banding method does not use ethidium bromide. Weigh a flask, and then add the lysed nuclei and weigh the combination. Add 1.262 times the lysate mass of CsCl, mix well, and centrifuge at 42,000 rpm as above. Collect fractions of the gradient, assay the

fractions by A_{260} or by viscosity, and pool the DNA fractions. Dialyse them against 2 liters of TE, pH 8, leaving a large space in the bag for expansion of the liquid. The DNA solution is very viscous, so dialysis takes a long time. Dialyse 3 times for 1 day each.

6. Phenol extraction may be performed at this point, but it can lead to lower molecular weight DNA unless it is done very gently. Extract the DNA until no interface occurs between the 2 phases. Then extract the aqueous phase 3 times with water-saturated chloroform. Reextract all of the phenol and chloroform phases with a single aliquot of about 20 ml of TE, pH 8. This can increase the yield by 10–30%. Combine the extracts, reextract, and dialyse against TE, pH 8, once again.
7. Most carbohydrate can be removed by high-speed centrifugation of the DNA when the DNA is at a concentration of less than 200 μg/ml. Centrifuge in a microfuge or at 13,000 rpm in the Sorvall HB4 rotor or some similar rotor.

RECIPES

1. Embryo buffer (10×). To make this solution weigh CAPS (see below) in a 1-liter beaker. Do not use weighing trays or paper because CAPS sticks to them. Then add the $CaCl_2$ and the hexylene glycol and stir. Bring this to a volume of 1 liter. Store the solution at room temperature because it precipitates in the cold. The embryo buffer is used at a 10-fold dilution, that is, at 1×, and its pH must be adjusted to 9.2 at that dilution. The diluted stock can be stored at 2 °C without any precipitation. The ingredients of the 10× embryo buffer are as follows:

   2 M hexylene glycol (2-methyl-2,4-pentanediol)
   10 mM $CaCl_2$
   200 mM CAPS (cyclohexylaminopropane sulfonic acid)

2. Pronase. This is made at 5 mg/ml in 10 mM Tris–HCl, pH 8, 1 mM EDTA, and is self-digested for 1 h at 37 °C. It is used the day it is made.
3. Lysis buffer. This is 5% Sarkosyl, 0.5 M EDTA, pH. 9.2. Sarkosyl is used instead of SDS because it is more soluble in the highly concentrated CsCl solution used in the centrifugation.

PREPARING NUCLEOSIDE TRIPHOSPHATE SOLUTIONS

Commercial triphosphate stocks contain variable amounts of water. In addition, the triphosphate often dissolves to give an acid solution owing to insufficient neutralization of the phosphates. Therefore, it is essential to check the concentration of the nucleoside by reading the absorbance of the solution and also to check the pH of the solution. Note that the pyrophosphate linkage is labile under acid conditions.

PROCEDURE

For a 100 mM solution:

1. Weigh out ~200 mg
2. Dissolve in ~1-1.5 ml H_2O
3. Neutralize to about pH 7 with 1 M NaOH (use indicator paper)
4. Read absorbance of a suitably dilute sample at the λ_{max} for pH 7
5. Typically 50-200 μl of the base is required
6. Adjust to 100 mM by dilution

Triphosphate	λ_{max} (nm) at pH 7-10	$A_{\lambda max}$ for 100 mM solution
rATP or dATP	259	1540
rCTP or dCTP*	272	910
rGTP or dGTP	252	1370
rUTP	262	1000
dTTP	267	960

COMMENTS

1. If really pure triphosphates are required, it is advisable to repurify the triphosphate from contaminating diphosphate by electrophoresis.
2. The proportion of contaminating mono- and diphosphate nucleosides can be assayed as described in the next section.

CHROMATOGRAPHIC ANALYSIS OF NUCLEOSIDES

This method separates mono-, di-, and triphosphoryl 5′-nucleosides and can be used for qualitative or quantitative estimation of the proportions of these compounds. It can also be used to separate and quantitate the incorporation of precursors into nucleic acids.

PROCEDURE

1. Cut a strip of polyethyleneimine (PEI; Polygram, CEL 300 PEI, Brinkman Instruments) paper to the dimensions of a

* Literature values for dCTP are for pH 2. We assume that the absorbance of dCTP at pH 7 is the same as rCTP at pH 7.

25 × 75 mm glass microscope slide. Then draw a light pencil line 0.5 cm from the bottom and along this line draw 3 light, equally spaced crosshatches. Do not break the matrix surface when drawing these light lines. Use the pencil to etch a line through the matrix of the paper about 0.3 cm from the top as shown in Figure 5.7. It is convenient to make many of these strips at one time and then store them at 4 °C for future use.

2. Use double-stick tape or a ring of tape with a single-stick surface to stick this strip, plastic side down, onto a lab bench. While blowing cold air over the strip with a hair dryer, spot, at the central crosshatch, 1 μl of a mixture of the appropriate mono-, di-, and triphosphoryl 5′-nucleoside standards, each 5 mM in dH_2O. Add this by briefly and repeatedly touching the paper with the pipette tip. Capillary action will draw the liquid onto the paper. Let the paper dry after adding each aliquot so that the liquid spot always remains less than about 0.2 cm in diameter. This will confine the nucleosides to a small area and will give good chromatographic resolution. The nucleoside standards should have the same base and sugar as the unknown. Spot approximately the same quantity of the unknown in an equal volume of dH_2O onto one of the other crosshatches. If the unknown is radiolabelled, it should be spotted on top of a spot of the nucleoside standards.
3. Make a 3.2% (wt/vol) solution of ammonium bicarbonate and put approximately 2.5 ml of it in the bottom of a Coplin Staining Jar. Place the PEI strip into the slide slots of the jar so that the sample end of the strip is in the solution. The volume of solution should not submerge the sample spots but should wet the entire bottom edge of the strip.
4. Cap the jar and wait 10–20 min until the solvent front has reached the etched line. Dry the strip with the hair dryer as before.
5. Detect the nucleosides by illuminating the strip with a standard hand-held short-wavelength UV light. The nucleoside areas are dark spots.
6. If quantitation is necessary, photograph the strip while illuminating with UV light. Determine the amount of radiolabel by circling the nucleoside spots with a pencil, using a razor blade to scrape the encircled areas into separate scintillation vials that contain 1 ml of the ammonium bicarbonate solution, shaking for a minute, and then adding an aqueous scintillation fluid and counting in the scintillation counter.

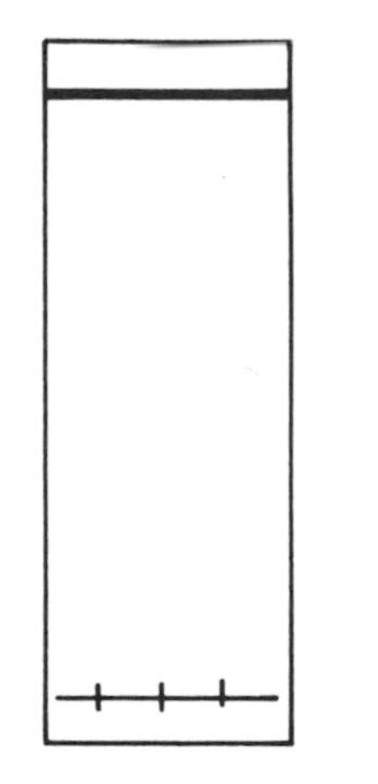

Figure 5.7. The use of PEI strips for separation of nucleotides.

Note: The order of migration in this buffer system is monophosphates first, diphosphates, and then triphosphates. Polymerized nucleic acid will not move.

GEL ELECTROPHORESIS OF DNA

At one point or another, virtually every nucleic acid experiment requires electrophoresis of a sample. The simple feature of being able to separate molecules according to size has made gels useful for a large number of purposes, ranging from assays of enzyme activity to fairly precise determinations of molecular lengths or the isolation of particular molecules from a population of molecules with different sizes. General features of gel electrophoresis are described by Helling et al. (1974), Maniatis et al. (1975), and McDonell et al. (1977). We describe only DNA electrophoresis. For short RNA we use the method of Maniatis et al. (1975) and for long RNA, the method of Rave et al. (1979).

Two DNA fragments that fall within the length range of one nucleotide to 50,000 base pairs can be resolved on gels provided that their length differs by at least 1%. This remarkably high resolution is one of the reasons gel electrophoresis is so useful. However, no single gel can separate molecules with such high resolution over this entire range of lengths. As might be expected, gels with more dense nets of fibers are more effective in separating smaller molecules. Although the useful separation ranges of various gels do overlap and the exact electrophoresis conditions can shift their useful range, the following is a helpful starting point for picking a gel for a particular purpose.

Gel	Separation range (bases or base pairs)
0.3% agarose	50,000 to 1000
0.7% agarose	20,000 to 300
1.4% agarose	6000 to 200
4% acrylamide	1000 to 100
10% acrylamide	500 to 25
20% acrylamide	50 to 1

For each of these gels, with the exception of the 20% acrylamide gel, it has been found that the logarithm of a molecule's length is proportional to its migration velocity over a broad range. For a typical gel this relationship can be plotted on semilog paper to give the curve shown in Figure 5.8. As a result, the length of an unknown fragment can be closely determined by coelectrophoresis with other fragments of known length. The known fragments allow construction of a calibration curve into which the unknown can be interpolated, yielding an accurate estimate of the unknown length. One should be cautious, however, in extrapolating the linear portion of the curve to either the high or the low molecular weight limits of a gel, because the linear relationship breaks down.

The limits as well as the slope of the linear portion of this plot are fairly reproducible. However, they depend on such

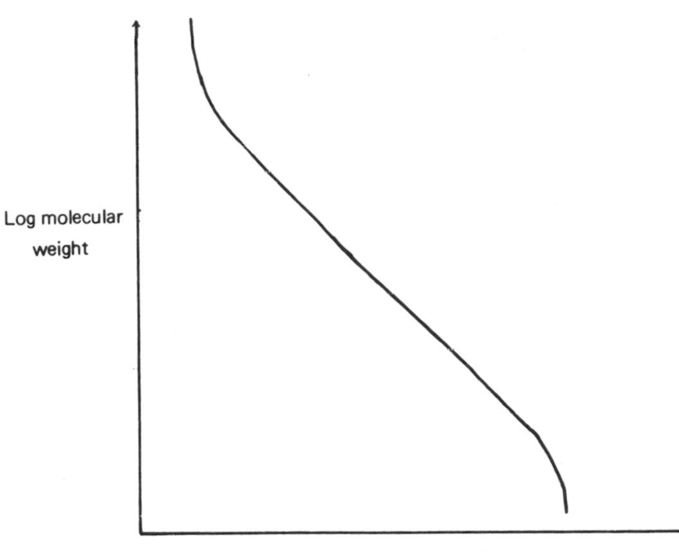

Figure 5.8. The typical relationship between molecular weight and migration velocity on agarose or acrylamide gels.

obvious variables as gel concentration, buffer, ionic strength, pH, and the voltage drop per distance along the gel, as well as on such less obvious variables as the temperature of the gel and the length of time the gel has been allowed to polymerize. As an important example, the useful size separation for higher molecular weights can be compromised by attempting to migrate the DNA too rapidly. A possible explanation for the loss of size dependence is that the DNA is migrating end-on through the gel.

A wide variety of buffers can be used for electrophoresis. It is rumored that chloride ions should be avoided in electrophoresis buffers because the electrode reaction produces sufficient chlorine and other ions that DNA is harmed.

Agarose Gel Electrophoresis

There are a wide variety of designs for the hardware of agarose gel electrophoresis and almost all of them work equally well. There are two fundamental differences in design. One is in the electrical connection between the buffer in the gel and in the reservoirs at the positive and negative poles. Some setups permit direct contact between the gel and the reservoir and others use a paper wick. The one described below uses paper wicks whose only apparent drawback is that a substantial portion of the voltage delivered by the power supply appears across the wicks. This should be remembered when trying to reproduce conditions in which a voltage gradient within a gel was specified. To avoid having to determine the actual voltage across a gel, we usually reproduce the current flow through a gel. It is rumored that direct contact between the gel and the buffer reservoir, as in submerged gels, will increase DNA migration rate. If the same gel, gel dimensions, electrophoresis buffer, and current flow are used, then the DNA migration rate should in theory be the same. In our experience it is the same.

The second fundamental design difference is between horizontal and vertical gels. The horizontal gel apparatus sketched below has several advantages, as follows: samples are easily loaded; the gel can be stained in the apparatus after electrophoresis; vacuum grease does not need to be used; DNA migration can be easily followed during electrophoresis; gel formation is very simple; and the apparatus is fairly inexpensive.

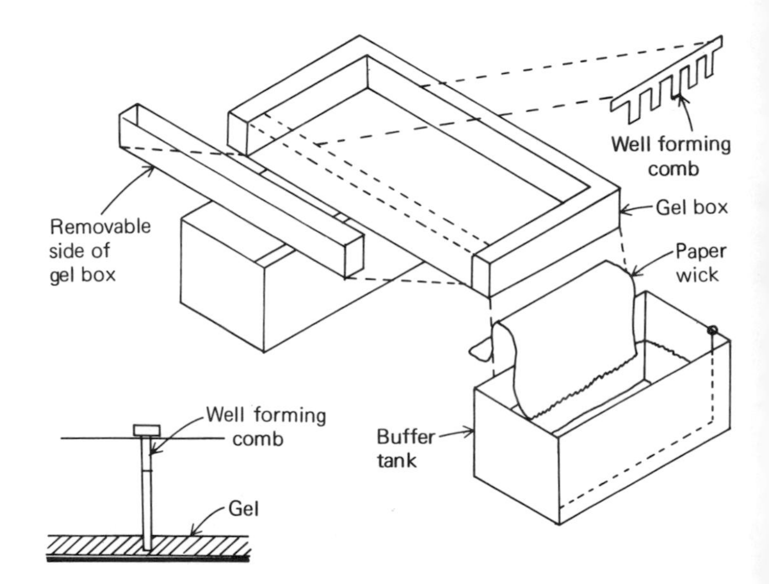

Figure 5.9. An exploded version of a Plexiglas, horizontal gel electrophoresis apparatus. The insert shows that the comb is made so as not to make contact with the bottom of the electrophoresis tray during formation of the wells. The gel box is 14 × 25 cm.

The gel box is made of Plexiglas and has one removable side, so that after electrophoresis the gel can be gently removed from the box. The dimensions shown (Figure 5.9) have proven convenient for routine work, but the simplicity and low cost of the design make it practical to construct boxes with different dimensions for specialized uses. Of course the gel width can be varied by using different positions for the movable side wall.

RECIPES

E buffer	10 × E buffer
40 mM Tris	96.9 g Tris base
20 mM Na^+-acetate	32.8 g Na^+-acetate
2 mM EDTA	15.2 g EDTA
18 mM NaCl	21.0 g NaCl
	H_2O to about 2 liters
	pH to 8.0 with acetic acid
	adjust to 2 liters

GEL FORMATION

1. Seal the removable side of the box with paper adhesive tape. This holds the side in place and prevents leakage of molten agar.
2. Place the comb in the box so that it is parallel to and about 3 cm from the end of the box. Pencil marks on the walls aid this positioning.
3. Melt the agarose in E buffer. A microwave oven is particularly convenient for this step, but an autoclave works also. Be sure to completely melt the granules of agarose, otherwise the resulting pattern of electrophoresed DNA will be useless. The apparatus described works well with 100 ml of agarose. Two or three times this amount can be used if the gels are being used preparatively. Agarose is available from several commercial sources. Sigma and SeaKem agarose work well, but we usually use SeaKem (ME; Marine Colloids, Inc., Rockland, Me.). The variety of grades available from Marine Colloids can be useful.
4. Allow the melted agarose solution to cool until the flask can be held in your hand without pain (about 55 °C); then pour it into the gel box and let it harden at room temperature for at least 1 h. Very low concentrations of agarose gels (below 0.5%) may require longer hardening periods and/or hardening at lower temperatures.
5. Remove the comb after the gel has hardened. Prevent damage to the wells by lifting one end of the comb until the first tooth is free of the gel and then holding that end of the comb fixed while lifting the other end of the comb. This minimizes the vacuum in the space between the gel wells and the comb.
6. Flood the gel surface with E buffer, making sure that all wells are filled. The gel may be used after a 10-min soaking. Gels may be stored for several days at room temperature and can be kept clean and wet by covering the box with a plastic sheet (Saran Wrap).

PROCEDURE FOR DOUBLE-STRANDED DNA

1. Fill each reservoir with about 300 ml of E buffer. More or less buffer can be used. If the buffer touches the solder connection between the platinum wire and the banana plug connector, electrode reactions will soon unsolder the connection. On the other hand, too little buffer in the tank will give poor buffering capacity and electrode reactions will change the pH drastically. In turn, this pH change can have dramatic effects on the hardness of the agarose. Finally, capillary action through the wicks can shift buffer from the reservoirs to or from the gel by a siphoning effect.
2. Pre-wet the wicks and place them so that approximately 1 cm of the wick rests on the end of the gel and the remainder takes the most direct route to the buffer tanks.

Do not let the wicks touch the paper tape used to seal the gel box. Some tapes contain dyes that will elute into the wicks and electrophorese into the gel. They fluoresce under UV illumination. Four layers of Whatman 3-MM paper can be used for wicks but such wicks must be covered with Saran Wrap to prevent excessive drying during electrophoresis. It is more convenient to use Telfa nonadherent strips (Kendall, Hospital Products Division, Chicago, Ill.). They have a plastic coating on one side which, if kept facing the air, prevents drying of the wick. Between uses, the wicks can be stored wet or dry. Keep the wicks from different poles separate and reuse them at the same pole to prevent contamination of one gel with material electroeluted from previous gels.

3. Prepare a sample for electrophoresis by adding 0.2 vol. of 5× sample solution (50% glycerol, 0.1% Bromphenol blue, 0.1% xylene cyanol, in 5× E buffer). The salt concentration in the sample should be less than 0.1 M to minimize smearing of the DNA bands. For an analytical 20-well gel of 100 ml, approximately 0.2 μg of DNA in a volume of 10–15 μl is added to a well. The glycerol allows the sample to settle to the well bottom. Larger sample volumes force some of the DNA to flow down the curve of the well meniscus and give bands that have a trailing smear (Figure 5.10). A 200-ml gel with only one large well can be loaded with 25 μg of DNA in 400 μl and will give

Figure 5.10. A gel well loaded to the maximum amount consistent with clean separation of bands.

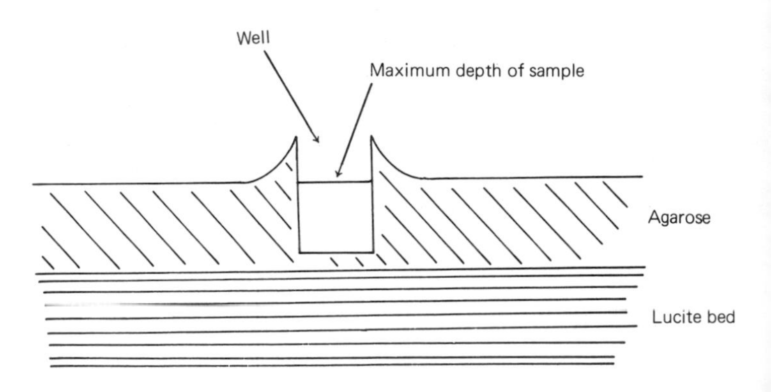

clear sharp bands and no overloading artifacts upon electrophoresis. The DNA quantities mentioned are nominal for DNA samples that will give less than 10 bands. If more bands are produced, then more DNA can be loaded on the gel without artifactual migration. If a very heterogeneous population of molecules, for example a restriction digest of DNA from the entire genome of *E. coli,* is to be used, then 2 μg of DNA can be loaded in a single well of a 20-well, 100-ml gel.

4. For a 100-ml gel, drive the samples into the gel with an 80-mA current for 15–30 min so that all of the dye has gone into the gel. Then turn the current off and distribute

about 5 ml of E buffer over the surface of the gel, and fill all wells so that they overflow. Put Saran Wrap over the entire surface of the gel making sure that no air bubbles are left in the wells. Bubbles in the wells will distort the electric field in the gel and may result in distorted band patterns.

5. For a 100-ml, 1% agarose gel, electrophoresis at 40 mA for about 10 h is adequate for most purposes. DNA migration rate can be manipulated by changing the ionic strength of the buffer or by changing the current. Lowering ionic strength or increasing the current will increase the migration rate. The rate of migration can be estimated by the migration of the bromphenol blue dye, which in such a gel comigrates with double-stranded DNA of about 0.3 kb. When connecting the electrodes, recall that DNA will migrate to the positive pole.
6. If too high a current is used the gel will heat up, apparently in a nonuniform way, usually causing DNA migration rates to vary in different parts of the gel.
7. It is sometimes necessary to follow the migration of bands during electrophoresis in order, for example, to maximize separation between two bands. This may be done by including ethidium bromide in the E buffer and by occasionally observing the DNA with a high-intensity short-wavelength UV light, such as Mineral Light Lamp Model UVS-54 (254 nm) (Ultra-Violet Products, Inc., San Gabriel, Calif.). This light is harmful to skin and eyes so use shielding.

PROCEDURE FOR SINGLE-STRANDED DNA

The procedure is largely the same as for double-stranded DNA. Agarose is melted in distilled water, cooled to 50 °C, and then 0.1 vol. of 10× NE (300 mM NaOH, 20 mM EDTA) is added and the solution is mixed thoroughly by swirling the flask. The solution will become slightly yellow. Pour the gel into the box and let it harden. Add to the sample 0.2 vol. of 5× sample solution without the 5× buffer and then transfer it to the gel well. Double-stranded DNA will denature in the wells. The migration rates of double- and single-stranded DNA of equal length are about the same in neutral and alkaline gels, respectively.

Occasionally, it is useful to electrophorese both double- and single-stranded DNA or RNA in different lanes of the same gel. We have done this following the denaturation and glyoxalization procedure of McMaster and Carmichael (1977) with the following modifications:

1. Oxidation products in the glyoxal may be removed simply by shaking the glyoxal with the mixed bed resin and pipetting the required glyoxal after the resin has settled.
2. Before staining, soak the gel in 50 mM NaOH and then neutralize in the sodium phosphate buffer described by

McMaster and Carmichael or in TE or E buffer. Stain the gels in 1 μg/ml ethidium bromide. The NaOH step will remove the glyoxal from the guanine residues and allow more efficient staining.

Polyacrylamide Gel Electrophoresis

Almost the same principles apply for electrophoresis of DNA in acrylamide as for electrophoresis in agarose. (For the properties of acrylamide and the assembly of gel sandwiches and reservoirs, see Polyacrylamide Gel Electrophoresis of Proteins, p. 80.) For the electrophoresis of DNA it is unnecessary that the top reservoir buffer make direct contact with the gel. The connection between the top of the gel and the reservoir can be made through a wick of Whatmann 3-MM paper. Another change from the protein gel procedure is that a longer time of gel polymerization is better. Although the gel sets in an hour, polymerization continues for at least 24 h. This additional polymerization has little effect on the migration of the DNA. However, unpolymerized acrylamide monomer has a high UV absorbance, so that using a young gel will drastically reduce the sensitivity when detecting DNA by ethidium bromide fluorescence.

Gels composed of less than 5% acrylamide are fragile. They can be toughened by including 0.5% agarose. This will not change the DNA migration rates.

RECIPES

1. TBE (10×)

Adjust to pH 8.3 by adding boric acid or Tris base and then adjust to 1 liter with H_2O.

0.9 M Tris	109 g Tris base
0.646 M boric acid	40 g boric acid
25 mM EDTA	7.3 g EDTA free acid
55.5 M H_2O	750 ml H_2O

2. Acrylamide-MBA

20% (wt/vol) Bio-Rad acrylamide, 0.67% MBA (methylenebisacrylamide). This can be stored for months in the refrigerator.

PROCEDURE

1. Prepare the gel sandwich as described in the section on protein gels. As in the case of protein gels, seal the sandwich with either high vacuum grease or paraffin wax. Make the gel solution described in the recipe section above using the procedure described for protein gels. After adding the TEMED, immediately pour the gel solu-

tion into the gel sandwich using a funnel or syringe. If agarose is to be added, melt a 1% agarose solution in TBE and, when it has cooled to approximately 55 °C, add an equal volume of the appropriate concentration of acrylamide gel solution prepared as above and then pour the gel immediately.

2. The gel can be "cured" for 24 h with or without the comb in place. Occasionally, if the gel has not been put in a cold room, curing for that length of time with the comb in place will allow some of the wells to dry out and will lead to distorted migration patterns. If the comb has been removed, seal the sandwich with Saran Wrap to prevent drying.
3. Remove the comb and rinse the wells with 1X TBE to remove unpolymerized acrylamide. Then assemble the reservoir apparatus and preelectrophorese the gel for at least 30 min before loading the samples. The samples can be in the same sample buffer as is used for agarose gels. As in the case of agarose gels, it is good not to have too high a salt concentration in the sample. Samples can be loaded with a disposable microliter pipette connected to a mechanical pipettor or a polypropylene plastic tube connected to a 50-μl Hamilton syringe. It is easiest to load samples by layering them under electrophoresis buffer.
4. Acrylamide gels separate DNA more quickly than agarose gels, and a run takes from 2 to 12 h. On a 5% acrylamide gel the dye markers xylene cyanol and bromophenol comigrate with DNA of about 260 and 65 base pairs, respectively. After electrophoresis the gel can be stained and photographed (see below) or, if the DNA is ^{32}P-labelled, dried (p.88) and autoradiographed (p. 184).

Staining and Photographing Gels

The most convenient stain for visualizing DNA in gels is ethidium bromide. (For properties of ethidium bromide staining, see Measuring DNA Concentration, p. 90.) Agarose and acrylamide gels can be stained by immersing them in about 400 ml of 1 μg/ml ethidium bromide. Stocks of ethidium bromide can be kept for months at 4 °C at concentrations between 100 μg/ml and 1 mg/ml. Several hours of staining are required for stain to completely penetrate a gel 5 mm thick and give the maximum sensitivity. However, 15 min of staining are sufficient to visualize as little as several ng of DNA in a band. It is rarely worth the effort to remove free ethidium bromide from a gel by washing it before photography. The agarose gels can conveniently be stained by adding 100–200 ml of staining solution to the gel while it is still in the electrophoresis apparatus.

A gel can be moved from the gel box to a photographic setup by sliding it onto a spatula that is the same size as the gel. The wall of a plastic dishpan can be carved into such a spatula. A gel can be released from its gel box by injecting water under the gel with a wash bottle. Lubricating the spatula and other surfaces with water will allow the gel to slide easily.

The source of UV light for exciting the fluorescence of ethidium is not critical. It is possible to use small hand-held lights and photographic exposures of several minutes. The most convenient sources of light are the transilluminators (Mineralight, Ultra-Violet Products, Inc., San Gabriel, Calif.). These provide an intense source of light through a filter about the size of a gel. The gel can be placed directly on the filter of the source and then observed or photographed from above. The short-wavelength sources (260 nm) provide about 5 times the sensitivity as the long-wavelength sources (365 nm), but the combination of ethidium bromide and short-wavelength UV light knocks the hell out of the DNA. Thus, if the DNA is to be extracted from the gel and used for some other purpose it is usually better to use long-wavelength illumination. Staining a strip from a gel and using this as a guide for cutting out the desired band of DNA avoids this problem.

Polaroid film is by far most practical for gel photography. Film types 107 and 667 have adequate sensitivity and resolution. A filter is provided with the Mineralight transilluminators which removes some of the purple light that interferes with observing the red-orange fluorescence of ethidium bromide. However, for maximum sensitivity, it is beneficial to use a red-orange filter between the gel and the camera. Be aware that too red a filter can substantially reduce the signal level. A 3-mm thick Corning filter CS-73 works well. A typical exposure for 10 ng of DNA in a band would be f-5.6 for 10 sec. Some Polaroid cameras will require inactivation of an interlock in order to permit manual timing of the longer exposures which are required for detecting fainter bands. When using the short-wavelength illumination of gels be careful to wear eye protection because the UV can cause a severe sunburn of the cornea.

Extracting DNA from Acrylamide and Agarose Gels

A multitude of methods have been used for extracting DNA from gels. None of them is entirely satisfactory, but the best in use at the present time are given below. If enzymes do not operate properly on gel-extracted DNA, they can usually be used successfully if bovine serum albumin is included in the reaction mix at a concentration of 250 μg/ml.

PROCEDURE FOR EXTRACTION FROM ACRYLAMIDE

The following elution method works very well for smaller DNA fragments. The recoveries begin to fall as the fragments become larger than several thousand base pairs. With the exception of extracting large DNA fragments the choice between this procedure and the electrophoretic method described later is largely a matter of taste.

1. Crush a 0.5-ml gel slice with a siliconized glass rod in an equal volume of 0.5 M NH_4^+-acetate, 10 mM Mg^{2+}-acetate, 1 mM EDTA, and 0.1% SDS. Overcrushing causes problems at later steps, so don't overdo it.
2. Incubate at 37 °C for at least 5 h and overnight if possible.
3. Separate the liquid from the fragments of acrylamide by centrifuging the slurry through a siliconized glass wool filter. This is conveniently done by poking the glass wool into the tip of a 1-ml Eppendorf pipette tip (Figure 5.11). This can then be put in a siliconized 10 × 75 mm test tube and the liquid centrifuged through the glass wool into the test tube by centrifuging at medium speed for several minutes in a clinical swinging-bucket centrifuge.
4. Reextract the acrylamide in the pipette tip by adding another 0.25 ml of the extraction buffer and centrifuging this into the centrifuge tube.
5. Do several cycles of ethanol precipitation to complete the removal of contaminants.
6. The recovery efficiency can be substantially increased by including 5 μg of carrier tRNA in steps 1 and 4.

Figure 5.11. The method for separating eluted DNA from agarose by centrifugation through an Eppendorf pipette tip.

PROCEDURE FOR EXTRACTION FROM AGAROSE

This is a simple and efficient method for extracting DNA from horizontal agarose gels. Its main drawback arises when the band containing the desired fragment is very close to an undesired DNA band. This problem can almost always be overcome by surgically removing the offensive band and then filling the newly created space with electrophoresis buffer. One major advantage of this method is that several DNA bands can be simultaneously extracted from the same gel. The extraction procedure described on page 124 is also satisfactory. A choice between the two is a matter of taste.

After the desired DNA fragment has been electrophoretically separated from neighboring bands, turn off the electric current and cut a trough in the gel just to the band's positive side. This can be done with a spatula while illuminating the gel with a long-wavelength hand-held UV light (Blak-Ray B100a, Ultra-Violet Products, Inc., San Gabriel, Calif.). Add 50% glycerol in electrophoresis buffer to the trough. Turn the current back on and continue electrophoresis until the fluorescence pattern indicates that approximately one-third of the DNA band has been eluted. Remove the liquid from the trough, refill the trough, and repeat the elution. After this has been done a third time, combine the elution fractions and add 0.05 vol. of 5 M NaCl and 2 vols of 95% ethanol. Then follow the ethanol precipitation protocol. After resuspending the DNA, centrifuge the DNA solution for 10 min in an Eppendorf microfuge to remove any large agarose fragments that remain. Virtually all of the DNA (95%) will be recovered in the 50% glycerol buffer.

An alternative method is satisfactory and more useful when fragments of considerably different electrophoretic mobility are being extracted from the same gel. This method will avoid the necessity of constantly attending to the gel. Follow the same procedure but cut a trough that not only is to the positive side of the DNA band but also extends for about 1 cm along the sides of the DNA band. This will form a "U" shape when looking at the upper surface of the gel when the gel is oriented so that its negative pole is at the top and the lower pole at the bottom. Place a strip of dialysis tubing so that it covers the bottom and the positive side of the trough and extends under the portion of the gel that contains the DNA band. This membrane will prevent the electroeluted DNA from entering the gel on the positive side of the trough. Add the 50% glycerol–electrophoresis buffer and electroelute until all DNA is out of the band. Reverse the current for 30 sec in order to elute DNA adhering to the dialysis membrane. Add salt and ethanol and precipitate the DNA. Resuspend the DNA and then remove large agarose particles with a 10-min centrifugation in an Eppendorf microfuge.

Note that this alternative method will elute the DNA into a smaller volume. This is often handy when many DNA samples are eluted and must be precipitated.

PROCEDURE FOR EXTRACTION FROM AGAROSE OR ACRYLAMIDE GEL SLICES

The DNA band can be detected in the gel either by ethidium bromide staining and illumination with long-wavelength UV light or by autoradiography. Remove the band by cutting the gel with a spatula. Place the gel slice inside a dialysis bag and add a small amount of a low conductivity buffer. We usually use 5 mM Tris, 5 mM boric acid, 0.1 mM EDTA (0.05 × TBE). Then place the bag, tied at both ends, into a shallow plastic box that contains two platinum electrodes running along opposite sides of the box. Arrange the bag so that it is between and parallel to the electrodes. Add 0.05 × TBE to the box in order to make electrical contact between the electrodes and the bag. Electrophoresis for 30 min at 200 V is usually sufficient to elute virtually all ($>$ 95%) of a DNA fragment smaller than 1 kb. Larger DNA will require longer electrophoresis or higher voltage. Remove the DNA solution from the bag after the power is turned off. If recovery is low, try adding tRNA carrier to the dialysis tube before electrophoresis. Also try reversing the current for 10 s after electroelution from the gel. The latter method will elute DNA that has been forced onto the surface of the dialysis bag. Remove large agarose or acrylamide particles by centrifugation in an Eppendorf microfuge for 10 min.

The same method can be done with one dish (9 × 5 cm) and two plastic buffer tanks from the agarose gel apparatus described above. Place the dish between the two boxes and put the bag in the dish in such a way that it parallels the electrodes in the two boxes. Add enough buffer to the dish and boxes; then use two paper wicks (4 layers of 3-MM paper or the Telfa pads) to connect the dish to each box. This apparatus will give more resistance because of the wicks. Usually 30 min at 1000 V is sufficient to transfer the DNA fragments smaller than 1 kb when 3-MM paper or Telfa pads are used for wicks.

MAPPING RESTRICTION ENDONUCLEASE SITES ON DNA

Although restriction enzyme reaction conditions are well described in the scientific literature and in enclosures provided with commercial enzyme shipments, the strategy for constructing maps is rarely, if ever, described. There is no one best strategy for all mapping ventures, but we have found a few elements of strategy that often simplify matters. In our description of these elements we assume that the DNA to be mapped is 5–50 kb in length, pure, and can be obtained in 10- to 50-μg quantities.

It is usually best to start by digesting separate aliquots of the DNA with different and inexpensive restriction enzymes that have six-base-pair recognition sites. The Xho I, EcoR I, Hind III, Kpn I, and Xba I enzymes are good candidates. Each of these makes infrequent cuts in most DNAs and gives DNA fragments with an average size of about 3–5 kb. Use each of these enzymes to digest different 2-μg aliquots of the DNA to be mapped. Analyze 0.2-μg samples of each aliquot by electrophoresis to determine if the digestion has gone to completion and to determine the number and size of the digestion products. If digestion of a sample is complete and if only a few restriction cuts have been made, 0.2-μg samples of the remaining 1.8 μg can then be digested with each of other enzymes that make only a few cuts in the DNA. Analysis of the size of these double digests can then be used to determine the relative locations of the two kinds of sites. In general, if the single and double digests of a particular pair of enzymes do not easily lead you to an unambiguous map, it is best to concentrate on other enzyme pairs that do not give ambiguous maps. As unambiguously located sites accumulate on the map, it will usually become obvious how a particular double or triple digest will clear up an ambiguity that has remained.

One advantage of this strategy is that the first digest of a double-digest pair has always been digested to completion. Incomplete digests are a major source of delay in deriving a map and some can be avoided by this strategy. Incomplete digests can usually be recognized by the pattern of fluorescence intensity in the ethidium bromide-stained DNA bands of a digest. Bands of lower molecular weight that have greater fluorescence intensity usually indicate partial digestion. They may also indicate that two different restriction fragments are the same size, so be careful.

It is usually best to perform the first digestion in a concentrated DNA reaction mix, so that second digests can be done simply by diluting an aliquot of the first digest into the appropriate second reaction cocktails. This strategy can not be carried on forever, so we present a table of restriction digestion conditions for many of the common enzymes (Table 5.1). This table will allow you to choose which enzyme reaction of a series should be done first, thereby enabling you to do the second reaction simply by adding one or two reagents. In general these conditions are for optimum enzyme activity. These optima are usually broad so that, for example, enzymes that require 10-50 mM Tris work well at any concentration in that range. Almost all enzymes will work well in 10 mM $MgCl_2$, 10 mM Tris, pH 7.5, 1 mM dithiothreitol, 50 μg/ml bovine serum albumin (BSA) at 37 °C provided they are supplemented with the proper amount of NaCl. If the ionic strength used is much lower than that specified in the table, at least two enzymes, EcoRI (Polisky et al., 1975), and BamHI can begin cutting at sites different from those specified in the literature. O'Farrell et al. (1980) have reported that a single buffer (33 mM Tris-acetate, pH 7.9, 66 mM K^+-acetate, 10 mM Mg^{2+}-acetate, 0.5 mM dithiothreitol and 100 μg/ml BSA) works well with almost all of the common restriction enzymes.

When constructing these maps it is important to use the same reliable molecular weight standard in all gels. If a switch is made from one molecular weight standard to another, be sure to compare the two standards in a single gel. Usually you will have to calculate and use a correction factor to compare sizes of fragments that were determined using different standards. For DNA in the 5-30 kb range, a Hind III digest of lambda DNA is a good choice for standards. For shorter DNA, a digest of pBR322 is probably the best choice. For the sizes of the fragments produced by several of the most useful digests, see Appendix II, Useful Numbers. For the sizes of other digests see Fiandt et al. (1977), Rosenvold and Honigman (1977), and Sutcliffe (1978). If you do use lambda DNA, be aware that the sticky ends of lambda can join to give you an additional fragment. Heating the lambda DNA at 65 °C in 10 mM Tris, 1 mM EDTA, pH 8 for 5 min immediately before loading on the gel will reduce but not eliminate this problem.

Enzymes that recognize only 4 bases usually make so many cuts in large DNA that the cutting sites cannot be mapped without adopting a divide and conquer strategy. Try isolating (p. 122) one restriction fragment from the other fragments of the starting DNA and then digest this fragment with the enzyme that recognizes 4 bases. End-labelling and partial digestion by a restriction enzyme (Smith and Birnstiel, 1976) is frequently the best way to map sites on such an isolated fragment.

Table 5.1. Restriction enzyme reaction conditions.

Enzyme	$MgCl_2$ (mM)	Tris (mM)	pH	NaCl (mM)	KCl (mM)	Dithiothreitol[a] (mM)	BSA (μg/ml)
Alu I	6	6	7.6	60	–	1	50
Ava I	6	6	7.4	60	–	1	50
BamH I	6	6	7.5	60	–	1	50
Bgl II	6	6	7.4	60	–	1	50
BstE II	6	6	7.9	60	–	1	50
EcoR	6	100	7.5	60	–	–[b]	50
Hae III	6	6	7.5	6 (or 60)	–	1	50
Hha I	6	6	7.4	60	–	1	50
Hind III	6	6	7.4	60	–	–[b]	50
Hinf I	6	6	7.4	60	–	1	50
Hpa I	6	6	7.4	60	–	1	50
Kpn I	6	6	7.5	6	–	1	50
Msp I(Hpa II)	6	6	7.4	60	–	1	50
Pst I	6	6	7.4	60	–	1	50
Pvu II	6	6	7.5	6 (or 60)	–	1	50
Sac I	6	6	7.4	–	–	1	50
Sal I	6	6	7.5	150	–	1	50
Sau3A I	6	6	7.5	60	–	1	50
Sma I	6	6	8.0	–	20[c]	1	50
Xba I	6	6	7.4	150	–	1	50
Xho I	6	6	7.4	150	–	1	50

[a]10 mM 2-mercaptoethanol can be substituted for 1 mM dithiothreitol.
[b]It does no harm to include dithiothreitol or 2-mercaptoethanol in these reactions.
[c]It may be possible to substitute NaCl for KCl in this reaction.

The reaction mixes can be conveniently made from the following sterile stock solutions:

60 mM $MgCl_2$, 60 mM Tris-HCl, pH 7.5, 10 mM dithiothreitol. Autoclave the Tris-$MgCl_2$ solution and when cool add the solid dithiothreitol and BSA.
1.5 M NaCl
600 mM NaCl
60 mM NaCl

An alternative is to make two solutions, a low salt and a high salt mix (see below), which at 10-fold dilution will make almost all of the reaction mixes. In our experience all the enzymes work well when the 10-fold concentrated salt mixes are added to DNA in TE, pH 8, and NaCl is added as suggested by the above table.

10× low salt mix: 60 mM $MgCl_2$, 60 mM Tris-HCl, pH 7.5, 60 mM NaCl, 10 mM dithiothreitol and 500 μg/ml BSA
10× high salt mix: same as 10× low salt mix, but use NaCl at 600 mM

Chapter 6

Constructing and Analyzing Recombinant DNA

This chapter describes a number of procedures commonly used to construct, select, and characterize recombinant DNA. They could, for example, be used to obtain a recombinant DNA clone that contains a particular gene. The procedures are listed in approximately the same order as for a gene cloning project. The exception to this order is that RNA purification methods are grouped at the end of the chapter. Ordering the procedures according to a gene cloning scheme is convenient but somewhat misleading, since the procedures described are useful for many other purposes.

Throughout this chapter we will use the word "vector" to mean the bacterial plasmid or phage DNA into which another DNA fragment, the "insert," has been inserted by in vitro methodology.

JOINING THE ENDS OF DNA MOLECULES

Joining two DNA ends together is the key operation when forming in vitro recombinants. There are basically three strategies for joining: (1) ligation of the sticky ends generated by restriction endonuclease digestion (Dugaiczyk et al., 1975); (2) ligation of blunt ends generated by, for example, shear breaking (Hogness and Simmons, 1964) and either digestion with the single-strand specific nuclease, SI (Seeburg et al., 1977), or treatment with *E. coli* DNA polymerase I (Seeburg et al., 1977); (3) addition of adapters and then joining the adapters together either by ligation or annealing to form stable joints. In the third strategy, ligation is necessary following the addition of commercially available artificial restriction site adapters (Heyneker et al., 1972; Maniatis et al., 1978) or following the addition of restriction fragment adapters that, by virtue of having two different restriction site ends, can convert one sticky end into another. Artificial homopolymer tails, a third kind of adapter, are made long enough that their annealed joints do not melt under physiological conditions and hence no ligation is necessary (Wensink et al., 1974).

Despite this variety of strategies for joining molecules, only a few enzymatic reactions are involved. In this chapter we describe the following: S1 digestion, ligation (blunt and sticky), lambda exonuclease digestion, and homopolymer tail formation. The other major class of reactions, restriction digestion reactions, was described in the previous chapter.

Making Blunt Ends by S1 Digestion

DNA ends generated by shear (Hogness and Simmons, 1964) or by restriction digestion are often poor substrates for subsequent joining reactions. The shear-generated ends are a mixture of 5′ and 3′ protruding single strands and blunt ends. The ends generated by a particular restriction enzyme are either 5′ or 3′ single strands or blunt ends. In order to make a more uniform substrate of the shear product or a better substrate of the restriction product, it is frequently necessary to digest them with the single-strand specific nuclease (S1) of *Aspergillus oryzae* (Wiegand et al., 1975). Digestion with S1 will make most ends blunt or almost blunt. Such ends are fairly good substrates for blunt-end ligation and terminal transferase tailing. For the latter reaction we prefer the substrate generated by lambda exonuclease (see below), but this exonuclease is not always available.

PROCEDURE

1. Resuspend ethanol-precipitated DNA in a 1.5-ml polypropylene microfuge tube to achieve a final concentration of 10 μg/ml in 200 mM NaCl, 30 mM Na^+-acetate, 5 mM $ZnSO_4$, pH 4.5.
2. Incubate at 20 °C for 5 min, then add 2000 units/ml of S1 and incubate for 30 min.
3. After the 30-min incubation, chill the reaction on ice and extract twice with an equal volume of water-saturated phenol at 4 °C. Then extract twice with chloroform and reextract the phenol and chloroform with a small volume of TE, pH 8. Combine the extracts and ethanol-precipitate the DNA.

Ligation with T4 DNA Ligase

A wide variety of conditions can be used to ligate the ends of DNA (Gellert, 1971; Sgaramella, et al., 1970). The following general procedures work in most cases.

RECIPES

1. Ligase buffer (10×)

(store frozen at − 20 °C)
300 mM Tris-HCl, pH 8.0*
70 mM $MgCl_2$
12 mM EDTA
2 mM ATP
100 mM dithiothreitol
500 μg/ml bovine albumin (crystallized; Pentex, Miles)

2. Reaction mixture

1.5 μl 10× ligase buffer
0.1 μg DNA, 5 kb in length, in TE, pH 8
Sterile distilled H_2O to 15 μl total volume

PROCEDURE

1. Make the reaction mixture in a 1.5-ml polypropylene microfuge tube on ice.
2. Add 0.01 units (Weiss et al., 1968) of T_4 DNA ligase (New England Biolabs) for cohesive-end ligation and 0.1 units for blunt-end ligation and incubate at 4 °C for 18–24 h.

COMMENTS

1. When ligating DNA restriction fragments to form recombinant molecules, the relative rates of monomolecular (joining together the two ends of one molecule) and bimolecular end-joining are usually important. The general features of these rates as a function of DNA concentration have been described by Davidson and Szybalski (1971) and in more practical detail by Dugaiczyk et al. (1975). In the most common cases, where the molecular population that is to be cloned is complex, these articles are most useful in estimating the kinds of product variation that will occur as the total DNA concentration is changed or as the ratio between the vector and the other DNA is changed. The articles usually cannot be used to predict the absolute concentrations and ratios that will optimize for a particular product, but they will give you a good estimate of the proper concentration range.
2. Two direct methods have been used to reduce the number of vector plasmids that contain no insert. One is described in this chapter (see Cycloserine Selection of Recombinant Plasmids). It selectively kills cells that contain these plasmids. The second method removes 5′-phos-

* The pH adjustments are made at room temperature unless otherwise noted.

phates from both ends of the vector DNA (Ullrich et al., 1977) so that the two ends cannot be joined by ligase.

Lambda Exonuclease Digestion to Yield Free 3′ Ends

The technique of adding homopolymer tails to DNA for cloning purposes works considerably better if all the DNA strands to which tails are to be added extend beyond their complementary strands. This is because the tail-forming enzyme, terminal transferase, adds much more rapidly to extended ends and will give a very heterogeneous population of tail lengths if it is adding tails to a mixed substrate population of extended and unextended ends. Some reaction conditions for the addition of homopolymer tails have been devised that decrease the specificity for adding to extended ends (e.g., Roychoudhury et al., 1976), but they simultaneously introduce another problem, namely, tails are added to nicks in the DNA. We have greatest success when we generate overhanging 3′ ends on all molecules by performing a limited digestion with lambda exonuclease. The general strategy of this reaction (Little et al., 1967) is to saturate the DNA ends with enzyme and then to digest at a low rate so as to partially degrade all 5′ ends. The reaction is run at low temperature because the enzyme digests rapidly at 37 °C. Also, the reaction is run at enzyme excess because the enzyme digests processively. The conditions we describe will remove approximately 30 bases from 5′ ends that are base paired or are extending only a few bases beyond the 3′ ends of a duplex.

PROCEDURE

1. To a DNA solution that is 3 mM in DNA ends and 10 mM Tris-HCl, pH 8, 1 mM EDTA, add 1 M K^+-glycinate, pH 9.5, to 67 mM, 0.1 M $MgCl_2$ to 5 mM, and 1 mg/ml bovine albumin (crystallized; Pentex, Miles) to 50 μg/ml.
2. Bring the reaction mixture to 0 °C in an ice water bath in the cold room. Also chill all pipettes to 4 °C because the temperature coefficient of lambda exonuclease activity is high.
3. Add 8 units (about 2 pmol) of enzyme for every picomole of DNA ends.
4. Halt the reaction after a 45-min incubation at 0 °C by adding an equal volume of water-saturated phenol, 0 °C. Extract with phenol twice and chloroform twice and reextract the phenol and chloroform with an equal volume of TE, pH 8. Combine the aqueous phases from the extraction and reextraction and ethanol precipitate the DNA. All operations are done in microfuge tubes. For subsequent terminal transferase reactions resuspend the DNA in 10

mM Tris-HCl, pH 8, 0.1 mM EDTA at a concentration of 20 nM in ends.

Addition of Homopolymer Tails

RECIPE

2× TT solution. This double-strength reaction solution is made by adding compounds to water in the order listed below, adjusting the pH, and then bringing the solution to the final volume.

0.2 M cacodylic acid; adjust to pH 7.0 with KOH
0.016 M $MgCl_2$
0.2 mM dithiothreitol
0.015 M KH_2PO_4
1.0 mM $CoCl_2$
300 μg/ml bovine albumin (crystallized; Pentex, Miles)

This solution may be stored at −20 °C for months to years without loss of effectiveness. When thawed there will be a precipitate that includes the lavender $CoCl_2$. The precipitate should be resuspended by vortexing before the solution is used.

PROCEDURE

Addition of $(dT)_n$ tails:

1. Vacuum dry an appropriate amount of ^{3}H-labelled dTTP in a 1.5-ml polypropylene microfuge tube. Usually 25 μCi will allow sufficient resolution in detecting addition of nucleotides, but go through the calculations for your particular conditions.
2. Resuspend labelled triphosphates by adding 100 μl 2× TT, 2 μl 1.0 mM dTTP, and 100 μl DNA. The reaction mixture is usually about 10 nM in 3′ DNA termini. The DNA may be in TE, pH 8, but try to minimize the EDTA concentration in the reaction by using a more concentrated DNA stock and bringing the reaction to volume with distilled water.
3. For unknown reasons, the reaction rate seems to be somewhat different in different reaction mixes. For this reason, when the length of tails must be controlled within narrow limits, it is wise to run a test reaction with a 15-μl aliquot of the above reaction mixture. This will give a reliable prediction of the incorporation rate for the subsequent main reaction.
4. Incubate the reaction mixture for 5 min at 37 °C and then add 1 unit of calf thymus terminal deoxynucleotidyl transferase for each picomole of 3′ termini. Incubate at 37 °C. If necessary, dilute the enzyme in 2× TT. Take 5-μl sam-

ples of the reaction at 15-min intervals and assay incorporation by the TCA precipitation method. Under the conditions described, an average of 100 bases will be added per terminus in 30–60 min. At the time that the reaction should be complete, put it on ice and assay the time samples. The reaction rate will be slowed by several orders of magnitude when the temperature is lowered to 4 °C. After a period of 1–2 h at 0–4 °C the reaction can be reinitiated at the former rate if it is returned to 37 °C. More enzyme is usually necessary if the reaction is kept cold for a longer period.

5. Terminate the reaction by extracting twice with an equal volume of water-saturated phenol at 4 °C. Then extract twice with chloroform and reextract the phenol and chloroform with 100 μl TE, pH 8. Combine the extracts and ethanol precipitate the DNA.

COMMENTS

1. A simple calculation of average number of bases added per 3′ end is the following:

$$\text{bases/end} = \frac{\text{nM triphosphates in reaction}}{\text{nM DNA ends in reaction}} \times \frac{\text{cpm precipitable in reaction sample}}{\text{cpm total in reaction sample}}$$

2. An average of 100 bases/end is adequate for most cloning purposes.
3. Most problems with this reaction are due to impurities in the transferase enzyme preparation. The most common contaminant is a nicking enzyme that generates nicks during the reaction, thereby creating unwanted substrates for the terminal transferase enzyme. Commercial preparations frequently have less transferase activity than claimed.
4. Poly-dA and poly-dT tails at about 1 nM may be efficiently annealed within a few hours at 42 °C in 0.1 M NaCl, TE, pH 8.

E. COLI TRANSFORMATION WITH PLASMID DNA

The transformation of *E. coli* by plasmid DNA is a routine technique for introducing plasmids, reconstructed plasmids, and phage DNA into *E. coli.* The method we describe is based on a procedure by Mandel and Higa (1970).

PROCEDURE

1. Grow an overnight culture of the HB101 strain of *E. coli* K-12 in 10 ml of LB broth (see Appendix I, Commonly

Used Recipes) at 37 °C. A portion of this overnight culture may be stored at 4 °C for several weeks and used as an inoculum for subsequent overnight cultures. Some investigators find that at an unpredictable time, usually measured in weeks, the stored culture will no longer give overnight cultures that are highly transformable. This may be due to overgrowth by contaminating bacteria that are less transformable. If this occurs, a new overnight culture from a single cell isolate will need to be grown.

2. Dilute the fresh overnight culture at least 20-fold into LB broth. A 50-ml culture in a 250-ml flask is convenient for most purposes. Grow at 37 °C with vigorous shaking, that is, a setting of 7 to 8 on a New Brunswick G76 shaker bath.
3. When the cell density reaches 5×10^8 cells/ml, centrifuge 40 ml in a sterile 50-ml polypropylene Sorvall tube at 7000 rpm, 5 min, 4 °C in an SS34 Sorvall rotor.
4. Discard the supernatant and resuspend the cell pellet by vortexing in 20 ml of sterile 0.05 M $CaCl_2$, 0 °C. Leave the tube in ice for 15 min and then recentrifuge as before.
5. Discard the supernatant and resuspend the cells by vortexing in 4 ml of 0.05 M $CaCl_2$, 0 °C. This cell suspension can be stored with little or no loss of transformation competence by adding sterile glycerol to 15% (vol/vol), freezing aliquots in 1.5-ml polypropylene microfuge tubes with a dry ice–ethanol bath, and storing at −70 °C. Thaw these cells in ice water.
6. Add approximately 0.2 ml of the cell suspension (3 drops from a 2-ml pipette) to a DNA sample tube that is in an ice water bath. This should be done shortly after the cells have been resuspended or thawed. Each DNA sample tube has 0.1 ml of 10 mM Tris-HCl, pH 7.0, 10 mM $CaCl_2$, and 10 mM $MgCl_2$. In our experience higher salt concentrations lower the efficiency of transformation.
7. Shake the rack of sample tubes in the ice bath to give good mixing of the solutions. Incubate the tubes for 25 min in ice water, then 2 min at 37 °C and 10 min on the lab bench (approximately 22 °C).
8. Add 1 ml of LB broth and shake for 30 min at 37 °C. Then either spread 0.2 ml directly onto selective plates or add 2.5 ml of LB top agar (0.7%) and pour onto selective plates. The former method allows more reliable replica plating because the colonies are of more uniform size. The latter method is easier if many plates must be made, but since cells in the agar will grow with poor aeration until they reach the agar surface, they will take longer to become visible. Wait 2 days for the poured plates to develop all of their colonies.

COMMENTS

1. Expect $3–5 \times 10^6$ transformants/μg of intact pMB9 or pBR322 DNA. This efficiency may be increased by modi-

fications we have not used, such as prolonged incubation at 0 °C in step 4 (Dagert and Ehrlich, 1979) or the substitution of $RuCl_2$ for $CaCl_2$ (Kushner, 1978).

2. We recommend the use of 10 μg/ml tetracycline to select for transformants. Higher concentrations will lower the transformation efficiency. This effect increases with the molecular weight of the plasmid.
3. It is usually wise to control for all sources of bacterial contamination in your solutions. Plate the solution of untransformed bacteria, the $CaCl_2$ solution, the DNA solutions, the LB broth, and the soft agar.
4. YT medium seems to work as well as LB in all the operations.
5. *E. coli* strains vary in their ability to be transformed. Most C600 strains are transformed efficiently. HB101 strains (and presumably other strains) that are from different laboratories, but are supposed to be identical, frequently have very different transformation frequencies.

STORING STRAINS THAT CONTAIN PLASMIDS

Bacteria tend to be cured of (lose) plasmids when stored in or on nutrient agar. This is particularly true with some recombinant plasmids. Bacteria and their plasmids can be recovered after years of frozen storage and after repeated freezing and thawing if the following procedure is used.

PROCEDURE

Grow an overnight culture of the cells in LB broth and then add an equal volume of 2× freezing medium (see Appendix I, Commonly Used Recipes). Freeze and store at −70 °C. Rapid freezing in liquid nitrogen and slow thawing at room temperature allow recovery of viable cells from a 100-μl culture after more than 15 cycles of freezing and thawing and after at least 5 years of storage. Cells can be grown in an equal-volume mixture of LB broth and 2× freezing medium and then can be frozen and stored at −70 °C with similar viability after many freeze–thaw cycles.

CYCLOSERINE SELECTION OF RECOMBINANT PLASMIDS

When using the plasmid pBR322 (Bolivar et al., 1977) or pBR325 (Bolivar, 1978) as a vector for recombinant DNA cloning, it is possible to use drug resistance to select transformants that contain plasmids with DNA inserts. In many cases construction of the recombinant plasmid containing the insert proceeds efficiently and such a large proportion of the transformants contain inserts that they may be identified merely by spot testing onto the appropriate plates. For example, inserting DNA into the BamH I site of pBR322 inactivates the tetracycline resistance gene so that cells containing these plasmids are resistant only to ampicillin. Spot testing 100 amp^r transformants would identify the desired tet^s transformants. In other cases the frequency of insertion of foreign DNA is so low that scoring by spot testing is too laborious. It is then better to select for cells that lack a drug resistance. This can be done by killing the cells which are resistant to the drug (Bolivar et al., 1977).

The method we describe is for selection of amp^r tet^s cells, that is, cells containing a plasmid with foreign DNA inserted into the BamH I site. It provides at least a 100-fold selection of such cells. Recently, Bochner et al. (1980) have described a direct selection for tet^s cells utilizing fusaric acid. Although we have not used this method, it promises to be useful for selecting recombinant plasmids.

PROCEDURE

1. Transform by the previously described procedure, but in the final step add 1.0 ml of LB broth, incubate at 37 °C for 30 min, and then add 1 ml of 50 μg/ml ampicillin in LB broth and incubate at 37 °C for 3 h.
2. Add tetracycline to 20 μg/ml and incubate for 45 min at 37 °C.
3. Add cycloserine to 150 μg/ml and incubate for 3 h at 37 °C.
4. Pellet the cells at 6000 rpm for 5 min in an SS34 Sorvall rotor, and then resuspend in 5 ml of distilled water.
5. Pellet the cells as before and resuspend in 1.3 ml LB broth with 50 μg/ml of ampicillin.
6. Incubate for 30 min at 37 °C and then add 2.5 ml LB broth, 0.7% agar, and pour onto an agar plate with 50 μg/ml ampicillin.

IN VITRO RADIOLABELLING OF DNA AND RNA

Most of the techniques used to analyze and identify nucleic acids require some kind of hybridization reaction that is assayed by following radioactively labelled RNA or DNA. By far the highest specific radioactivities in nucleic acids are achieved by in vitro methods and these high activities lead to the most sensitive assays. We describe several of the most common and most reliable in vitro labelling methods.

Radiolabelling DNA by Nick Translation

This method utilizes the ability of *E. coli* DNA polymerase I (Kelly et al., 1970) to translate, or move, a nick. The polymerase does this by binding to a nick in the DNA, excising the nucleotide to the 5′ side of the nick, and then, while moving down the DNA strand, replacing that nucleotide using a triphosphate from the reaction mixture. By a series of these reaction steps the polymerase moves the nick down the DNA strand, replacing nucleotides from the DNA with radioactive bases from the triphosphates included in the reaction. The technique is somewhat delicate in that too little nicking of the DNA leads to inefficient incorporation of label, whereas too much nicking leads to a DNA product that is too short for most experiments. The basic strategy and conditions are described by Rigby et al. (1977).

The conditions described below will yield [^{32}P]-labelled DNA with a specific activity of approximately 10^8 cpm/μg or [^{3}H]-labelled DNA with approximately 10^7 cpm/μg. Iodine-labelled triphosphates are now commercially available (New England Nuclear) and are reported to give specific activities of 10^9 cpm/μg with the procedure described below.

The labelled product has an average single-strand size of approximately 600 nucleotides and is satisfactory for filter and solution hybridization reactions. Higher molecular weight products with somewhat lower specific activities can be obtained by lowering the DNase concentration 10-fold as long as the DNA polymerase I is not contaminated by significant endonucleolytic activity.

The procedure given is for [^{32}P]-labelling of DNA. For [^{3}H]-labelling, use labelled dCTP and dTTP for reasons of cost and specific activity and change the mixture of unlabelled triphosphates accordingly.

RECIPES

1. *Nick translation buffer* (10×), is made by mixing 2.5 ml of 1 M Tris, pH 7.8, and 0.45 ml of 1 M $MgCl_2$ and autoclaving. Then add 40 μl of 98% 2-mercaptoethanol (Eastman), 2.5 mg of albumin, and finally sterile water to yield a volume of 5.0 ml. This stock solution can be stored, tightly capped, at 4 °C for many months without loss of effectiveness. Freezing will give a precipitate. The final concentrations of the components in this buffer are as follows:
   500 mM Tris-HCl, pH 7.8
   90 mM $MgCl_2$
   100 mM 2-mercaptoethanol
   500 μg/ml bovine albumin (crystallized; Pentex, Miles)
2. *Stop buffer*
   10 mM EDTA
   0.5% SDS
   10 mM Tris-HCl, pH 7.4
   0.2 M NaCl

PROCEDURE

1. Vacuum-dry 0.2 nmol or 30 μCi of alpha-[^{32}P]-dCTP (> 300 Ci/mmol; Amersham or New England Nuclear) in a siliconized 13 × 100 mm glass tube or a 1.5-ml polypropylene microfuge tube.
2. Add the following solutions and bring the volume to 50 μl with distilled water.
   5.0 μl 10× nick translation buffer
   4.0 μl 100 μM dATP, dGTP, and dTTP. This yields 0.4 nmol of each unlabelled triphosphate in the final reaction mixture.
   0.2 μg DNA. This gives approximately 0.1 nmol of each nucleotide in a 50% G + C DNA. Up to 15 μl DNA in TE, pH 8, have been added without inhibition of the reaction.
3. Dilute DNase stock into 1× nick translation buffer at 4 °C. The DNase stock, 1.0 mg/ml (Worthington; DPFF), is stored in 10-μl aliquots at −20 °C in microfuge tubes. The appropriate dilution is crucial to the size and specific activity of the product. It must be determined empirically for each batch of frozen aliquots and for each lot of DNA polymerase. The latter is important because endonucleolytic activity frequently contaminates commercial DNA polymerase I preparations. If the polymerase has very little endonuclease and if the DNase is the usual Worthington product, a 1:46,000 dilution is fine. Start the dilution by adding 1 ml of 1× nick translation buffer, 4 °C to the 10-μl frozen aliquot of DNase.
4. Incubate the reaction mixture at 15 °C for 10 min and then add 1 unit of DNApolymerase I (Boehringer or Bethesda Research Labs) and 4 μl diluted DNase I. At 90 min store the reaction on ice and wait for the assay results.
5. Assay the reaction at 0, 30, 60, and 90 min by trichloroacetic acid (TCA) precipitation (p. 96) of approximately 0.1 μl of the reaction mixture. Add the sample to 200 μl of 2% SDS, 10 mM EDTA, 400 μg/ml BSA. Mix and then spot 20 μl of the mixture onto a GF/C glass filter (Whatman) as a measure of the total counts in the sample. Then add 2 ml of 5% TCA and follow the TCA precipitation procedure. The reaction should plateau by 60–90 min. At the plateau, between 20 and 90% of the radiolabel should be incorporated. Occasionally the addition of more polymerase will increase incorporation to a new plateau, but usually this is not worth the effort.
6. Stop the reaction by adding 150 μl of stop buffer and then add 15 μg of carrier DNA. Pass the reaction mix through a Sephadex G-100 column in TE to remove unincorporated triphosphates, the salts, and other contaminants. The 8-ml Bio-Rad Econo columns are convenient for this. Collect 0.5-ml (20-drop) fractions with a Gilson

Micro Fraction collector. If the fractions are collected in a convenient 72-hole polypropylene Gilson tray, the profile of radioactivity eluted can be assayed by a hand monitor using the homemade lead shield as shown in Figure 6.1. Make the shield by punching a 1-cm hole in a thin sheet of lead.

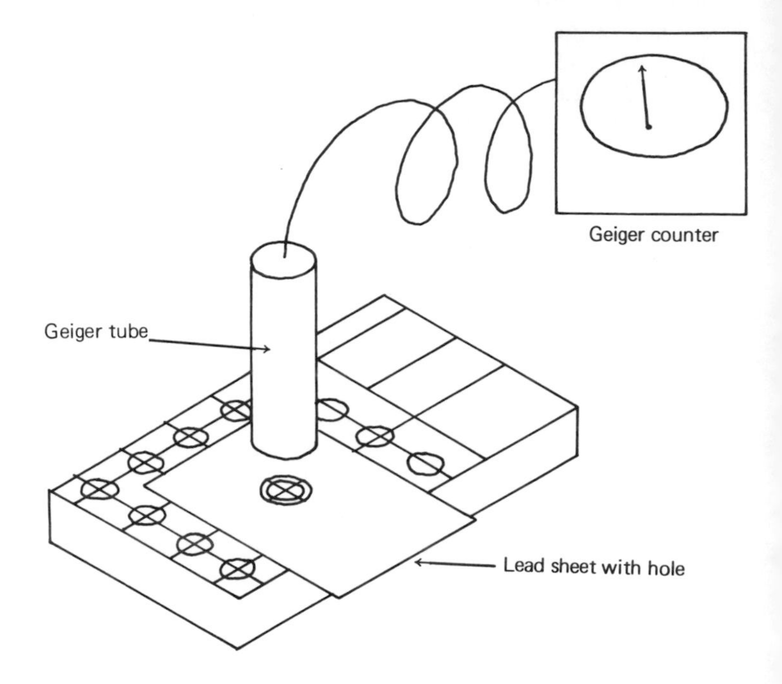

Figure 6.1. Determining or locating a radioactive fraction with a Geiger counter and lead shield.

Synthesizing Radiolabelled RNA Complementary to DNA

It is frequently useful to synthesize highly radioactive RNA complementary to a DNA sample. This is usually done so that the labelled RNA preparation can be hybridized to DNA and then the unhybridized RNA can be efficiently and selectively removed. RNase will digest the unhybridized RNA but will digest very little if any of the RNA that is in a duplex.

The radioactive RNA copy of DNA can be easily synthesized using *E. coli* RNA polymerase (Burgess and Travers, 1971) to nonspecifically transcribe the DNA into RNA. The method described below will yield [^{3}H]-labelled complementary RNA with a specific activity of approximately 10^7 cpm/μg. [^{32}P]-labelled precursors will yield specific activities about 10-fold higher.

PROCEDURE

1. Vacuum-dry the following in a siliconized 13 × 100 mm glass tube or a 1.5-ml polypropylene microfuge tube: 0.6 nmol of [^{3}H]-rUTP and [^{3}H]-rCTP (New England Nuclear;

about 40 and 25 Ci/mmol, respectively) and 15 nmol of rATP and rGTP.

2. Ethanol-precipitate 0.5 μg of DNA and resuspend in 50 μl of 1× nick translation buffer (p. 000) plus 5 μl of 1.5 M KCl.
3. Resuspend the nucleotide triphosphates in 50 μl of the DNA solution and incubate at 37 °C for 5 min. Add 2 units of *E. coli* RNA polymerase (Boehringer). Assay the reaction at 0 min and at 40-min intervals thereafter. After the 160-min sample, let the reaction continue and use the TCA precipitation method (p. 96) to assay the extent of RNA synthesis.
4. The incorporation of label into RNA should be linear for several hours. When incorporation has stopped, add 50 μg of carrier RNA and cool to 25 °C. Add 1 μg of DNase I (RNase-free; Worthington DPFF, the same as used in nick translation) and incubate at 25 °C for 30 min.
5. Add 50 μl of TE, pH 8, phenol extract twice, and reextract the 2 phenol phases with 100 μl TE, pH 8. Remove the aqueous phases and reextract the phenol phase a second time. Combine all 3 aqueous phases from the extractions.
6. Wash a 7.5-ml bed volume Sephadex G-100 column with 20 ml of TE, pH 8, 0.5% (vol/vol) diethylpyrocarbonate (Sigma). Load the aqueous phase extract on the column and collect the excluded volume as described for nick translation.

Synthesizing Radiolabelled DNA Complementary to RNA

In some experiments it is useful to synthesize radioactive DNA complementary to an RNA. This can be done utilizing the poly-A tail which is present on most eukaryotic RNAs. The RNA is first annealed to oligo-dT which then serves as a primer for elongation by avian myeloblastosis virus (AMV) reverse transcriptase (Kacian et al., 1971). By providing the reverse transcriptase with highly labelled triphosphates the DNA can be radiolabelled to a very high specific activity. One useful feature of this reaction is that the common contaminant of mRNA preparations, rRNA, is a very poor substrate because it lacks a poly-A tail.

RECIPES

Make all of these solutions from sterile water and sterile concentrated stock solutions of salts and buffers. We do not sterilize the commercial dithiothreitol or the oligo-dT, bovine albumin, or SDS.

1. *Salts* (5×), store frozen at −20 °C in 20-μl aliquots. Do not refreeze.

250 mM Tris-HCl, pH 8.2
50 mM $MgCl_2$
500 mM KCl
2 mM dithiothreitol

2. *Primer* (10×). Store frozen.
   100 μg/ml oligo-dT_{12-18} (Collaborative Research)
3. *Triphosphates* (5×), 5 mM each of dGTP, dATP, and dTTP (p. 111).
4. *Assay buffer*
   400 μg/ml bovine albumin (Miles; Pentex, crystallized)
   10 mM EDTA
   2% SDS
5. *Stop buffer*
   10 mM EDTA
   0.1% SDS
   0.2 M Na^+-acetate, pH 5.0

PROCEDURE

1. Vacuum-dry 0.5 nmol of alpha-[^{32}P]-dCTP in a 1.5-ml polypropylene microfuge tube or in a siliconized 13 × 100 mm glass tube.
2. Add the following while the tube is on ice:
   11.5 μl sterile distilled water
   5 μl 5× salts
   0.5 μl 200 mM sodium pyrophosphate
   2.5 μl 10× primer
   5 μl 5× triphosphates
   0.5 μl of 1 mg/ml poly-A^+ RNA
3. Incubate at 42 °C for 5 min and then add a saturating amount of AMV reverse transcriptase and incubate for 1 h at 42 °C. The enzyme is commercially available (Bethesda Research Labs), but we have no experience with the commercial product. The saturating amount must be determined empirically for each batch of enzyme. Saturating quantities of enzyme give longer transcripts than do subsaturating amounts. The quantity of primer oligo-dT is sufficient to saturate all the poly-A template in the reaction.
4. Take assay samples of the reaction at 0, 20, 40, and 60 min by transferring 0.5-μl samples to 200 μl of assay buffer on ice. At 1 h put the reaction tube on ice and follow the standard TCA precipitation protocol, (p. 96) to assay the incorporation. The reaction usually is complete at 1 h. Addition of more enzyme rarely will cause additional incorporation, whereas addition of more RNA will occasionally lead to more incorporation. Approximately 20% of the label will be incorporated when the reaction has stopped.
5. Add 200 μl of stop buffer, 30 μg of carrier tRNA, 200 μl of 5 M NH_4^+-acetate, and 3 vols of 95% ethanol, −20 °C. Store overnight at −20 °C.
6. Centrifuge for 10 min at 10,000 rpm in the Sorvall HB4 or

SS34 rotor. Carefully discard the highly radioactive supernatant. Resuspend the pellet in 300 μl TE, pH 8, and add 10 μl of 10 M NaOH. Incubate at 100 °C for 2 min to hydrolyse the RNA and then transfer to ice. Neutralize by adding 9 μl of 10 M HCl and then small volumes of 2 M HCl until the pH is 7. Vortex between additions of acid and measure the pH by spotting 1 μl on pH indicator sticks (EM Laboratories, Inc., N.Y.; or E. Merck, Darmstadt, Germany).

7. Add 30 μl of carrier tRNA and 2 vols of 95% ethanol, −20 °C, mix, and store overnight at −20 °C.
8. Pellet the DNA as in step 5, wash the pellet by swirling with 500 μl of 70% ethanol, −20 °C, and centrifuge. Discard the supernatant, dry the walls of the tube with a cotton swab (Q-tip), and vacuum dry the pellet. Resuspend the pellet in 200 μl TE, pH 8. Assay the fraction of label that is incorporated into the DNA by TCA precipitation.

Radiolabelling the 5′ Ends of RNA or DNA

Both RNA and DNA can be efficiently labelled by replacing their terminal 5′ phosphates with $^{32}PO_4$. The 5′ ends of single-stranded structures and the protruding, flush, or indented 5′ ends of double-stranded structures can be labelled. This method of labelling is useful for general purposes, but is particularly valuable when the label must be at the end of molecules. Such end-labelling is required in some sequencing procedures, in the Smith and Birnstiel (1976) method of restriction site mapping, and when nucleic acids, for example 5S RNA and tRNA, are difficult to label by other in vitro methods. The procedure is derived from Maxam and Gilbert (1977).

PROCEDURE AND COMMENTS

1. The specific activities obtained by this method are, of course, directly related to the number of 5′ ends and therefore to the size of the nucleic acid. For this reason higher molecular weight RNA is usually hydrolysed under conditions that yield 5′ hydroxyl ends on fragments that have an average length of approximately 50 to 100 nucleotides. This is done by incubating at 90 °C for 30 min in 5 mM glycine, pH 9.5, 10 μM EDTA, and 100 μM spermidine in a 1.5-ml polypropylene microfuge tube.
2. If flush-ended or 5′-indented dephosphorylated (see steps 5 and 6 for dephosphorylation) DNA is to be labelled, denature it by heating at 100 °C for 10 min in the same buffer. If the DNA to be labelled has less than a few thousand nucleotides of sequence information (complexity) and the amount of DNA to be labelled is more than 1 μg, then the DNA can renature and thus prevent efficient kinasing. To reduce this renaturation, quench the denatured

DNA on ice and immediately do the kinase reaction. Duplex kinased DNA is frequently the desired end product. The denatured DNA usually reassociates to form duplexes during the kinase and subsequent reactions of most procedures. To be sure that it reassociates introduce a hybridization step according to the description on p. 145.

3. Begin the kinase reaction by adding 45 μl of the RNA or DNA (1–50 pmol of ends) to γ-[^{32}P]ATP that is in molar excess to the number of 5′ ends and never less than 1 μM. The nucleic acid solution should not contain ammonium ion because even small amounts of this ion will inactivate the enzyme. If necessary, ethanol precipitate the nucleic acid and wash the pellet with 95% ethanol. Add 5 μl of 100 mM $MgCl_2$, 50 mM dithiothreitol, and 500 mM Tris-HCl, pH 9.5. Add 2 units of T4 polynucleotide kinase (New England Biolabs) (Richardson, 1971) and incubate at 37 °C for 30 min. Assay incorporation by TCA precipitation.
4. Remove unincorporated label by chromatography on DEAE-cellulose (DE52 Whatman; p. 94), by Sephadex gel filtration, or by three ethanol precipitations. In general, we prefer to remove unincorporated label by several ethanol precipitation steps done in the microfuge reaction tube. This method is rapid and limits our exposure to ^{32}P. The first precipitation of the nucleic acid is done in ammonium acetate solution instead of NaCl solution because the ammonium ion inactivates the enzyme. Add 200 μl 2.5 M ammonium acetate, mix, and add 1 μl 1 mg/ml tRNA and 750 μl 95% ethanol. Mix the tube by inverting several times, then put the tube in a dry ice–ethanol bath for 15 min. Centrifuge the tube for 15 min in an Eppendorf microfuge. After this remove the supernatant with a Pasteur pipette and resuspend the DNA in 250 μl of 0.25 M NaCl. Add 500 μl 95% ethanol, chill in dry ice–ethanol for 15 min, and centrifuge for 15 min in the Eppendorf microfuge. Wash the DNA pellet to remove any remaining salt. Do this wash by adding 1 ml of 95% ethanol, inverting the tube several times, and recentrifuging.
5. DNA that has phosphate on its 5′ ends must also be dephosphorylated prior to the kinase reaction. The procedure described for tRNA dephosphorylation (see step 6) works well for single-stranded DNA or for double-stranded DNA with protruding, indented, or flush 5′ ends. For single-stranded DNA or for 5′ protruding-end double-stranded DNA the dephosphorylation reaction can be done at 37 °C.
6. Because of its secondary structure tRNA is not reproducibly labelled by the above method and therefore should first be treated with bacterial alkaline phosphatase (BAP) (Worthington; BAPF). This treatment dephosphorylates the 5′ ends of RNA prior to its phosphorylation by kinase.

The BAP is dissolved at 0.5 units/ml in 0.1 M Tris, pH 8.3, and is dialysed against this buffer to completely remove ammonium ions which inhibit kinase. Dialysed BAP can be stored at 4 °C for at least 5 months without loss of activity. Add 10^{-4} units of enzyme per picomole of 5′ ends to the nucleic acid in 0.1 M Tris, pH 8.3, 60 °C and incubate for 30 min. The reaction conditions are not very critical; thus, the BAP reaction for DNA can usually be done immediately following restriction digestions simply by bringing the pH to 8.3 by adding Tris. This BAP reaction removes about 95% of the terminal 5′ phosphates. After the reaction, extract twice with water-saturated phenol and once with chloroform and then ethanol-precipitate (p. 130).

GENERAL ASPECTS OF NUCLEIC ACID HYBRIDIZATION REACTIONS

RNA-DNA and DNA-DNA hybridization reactions are the basis of many assays in recombinant DNA technology. In these reactions two single-stranded molecules anneal to form a base-paired duplex. Hybridization may be a misleading term to describe this reaction, but it is the one most commonly used.

The highest rate of nucleic acid hybridization occurs approximately 25 °C below the temperature, T_m, at which the DNA is 50% melted (denatured, strand-separated) (Studier, 1969). For this reason we generally hybridize at a temperature 25 °C less than T_m. When performing filter hybridizations, we follow hybridization with washes at (T_m − 15) °C in order to select for the more stable hybrids. It is important to bear in mind that the T_m varies with the G + C content of the nucleic acid and with the fidelity of the base pairing. Thus, any choice of hybridization conditions will cause some hybrids to be preferentially formed out of all the hybrids possible in a complex mixture of nucleic acids.

The 4 most useful relationships to consider when choosing hybridization and washing conditions are as follows:

1. $T_m = 69.3 + 0.41\,(G + C)\%$ (Marmur and Doty, 1962)
2. $(T_m)_{\mu_2} - (T_m)_{\mu_1} = 18.5 \log_{10}(\mu_2/\mu_1)$, where μ_1 and μ_2 are the respective ionic strengths of two solutions (Dove and Davidson, 1962)
3. A 1% increase in the number of mismatched base pairs decreases the T_m of a duplex by 1 °C (Bonner et al., 1973)
4. 1% Formamide reduces the T_m by 0.7 °C (McConaughy et al., 1969; Hutton, 1977)

Most of these relationships are approximate and were determined with SSC as the solvent. This solvent is a weak buffer and a weak chelator of divalent cations. At concentrations at or below 0.1× SSC, this solvent should be supplemented with 1 mM EDTA and 10 mM Tris to avoid major artifacts of nucleic acid stability due to trace amounts of divalent cations or unusual pH. If at all possible, avoid concentrations below 0.1× SSC.

For DNA with an average T_m of 85 °C in 1× SSC we hybridize at 65 °C in 2× SSC at 65 °C or in 4× SSC, 50% formamide at 37 °C (both are T_m − 25 °C); and for filter hybridization, wash in 0.5× SSC, 65 °C, or in 1.5× SSC, 50% formamide at 37 °C (both are T_m − 14 °C). The pH of the hybridization solutions is 7.5. Fifty percent formamide allows lower hybridization temperatures and is generally used for RNA-DNA hybridization because it gives a slower rate of RNA degradation. Note that many commercial preparations of formamide contain contaminating salts and formamide hydrolysis products. These can seriously affect the pH and the ionic strength of the hybridization mixture and should be removed (see Appendix I, Commonly Used Recipes).

DNA and RNA that are less than 50 bases long will be less thermostable (Laird et al., 1969). This will affect the rate of hybridization as well as the stability of the duplex. Most reactions we describe use short (50–500 nucleotides) radiolabelled nucleic acids to hybridize with other nucleic acids. In this range the effect of length on rate can usually be ignored. If both single-stranded species contributing to the new duplex are much longer, then the rate effect of length (Wetmur and Davidson, 1968) should be considered.

SCREENING RECOMBINANT DNA CLONES BY NUCLEIC ACID HYBRIDIZATION

In the course of recombinant DNA cloning projects it is common to produce many clones, only a few of which have the sought-after DNA segment. If, as is almost always the case, this segment does not have an effect on the host organism that can be selected by genetic methods, then the desired clone usually must be identified by nucleic acid hybridization techniques. For example, if a particular gene is to be purified from a shotgun collection of DNA segments cloned from an entire genome, it may be possible to identify it by using a radiolabelled, partially purified mRNA. To accomplish this identification a nitrocellulose paper replica of clones would be made in such a way that the DNA in the clones is denatured and immobilized on the paper at the location of clone growth. Then the radiolabelled mRNA would be hybridized to the paper and the clones containing complementary DNA would be identified by autoradiography. This general method was devised by Grunstein and Hogness (1975) for screening plasmid-containing colonies.

We describe below a variation that allows screening of phage plaques. We also describe a somewhat different method for screening colonies. It is very simple and has some other properties that frequently make it the method of choice.

Screening Phage Plaques

The recombinant phage are plated and allowed to develop until the plaques are almost confluent. Some of the phage from each plaque are then transferred to filter paper and the DNA in them is denatured and immobilized on the paper. This DNA is then allowed to hybridize with a [^{32}P]-labelled nucleic acid and, after washing, the location of hybridization is detected by autoradiography. Phage from the positive plaques can be recovered from the plaques on the original plate. This method was devised by Benton and Davis (1977).

SCREENING PROCEDURE

1. Grow 50 ml of an appropriate bacterial host in LB broth plus 0.2% maltose overnight with vigorous shaking. The A_{600} should reach approximately 2.
2. Pellet 40 ml of the cells in a sterile 50-ml polypropylene tube for 10 min at 5000 rpm in the SS34 Sorvall rotor. Resuspend the cells in 20 ml of 10 mM $MgCl_2$ at 4 °C.
3. When screening large numbers of phage, it is most convenient to screen them at the highest manageable concentration. Adsorb the phage to cells at 37 °C for 15 min in the proportion of 5 × 10^4 phage/1.75 ml of cells. Dilute 1.75 ml of this with 30 ml of 0.7% (wt/vol) agarose (not agar) in LB broth at 47 °C and pour onto 3 LB-agar petri plates of 140 mm diameter or onto 10 standard 90-mm diameter petri plates. If only 10^4 or fewer phage must be screened it is more convenient to adsorb at the same concentration and then spread at a greater dilution so that the plaques will be well separated. This can be done by diluting with a much larger volume of LB-agarose and then spreading on more plates.
4. Let the agarose harden for 10 min at room temperature and then incubate the plates at 37 °C, face up, until the plaques are nearly confluent. On high-density plates this usually requires about 6 h and on low-density plates 8–12 h.
5. Incubate the plates for at least 1 h at 4 °C and then start the phage transfer by placing a circle of dry nitrocellulose filter paper (Millipore; HA) onto the agarose surface of a plate so that no air bubbles are trapped between the agarose and the paper. Do this by holding the circle with tweezers at two points 180 ° apart and allow the middle of the circle to curve down onto the agarose. Do not allow the plates to warm up before transfer. If you have

to work at room temperature, then use the plates immediately after removing them from 4 °C. The filters should be marked for orientation by dipping a 22-gauge needle into India ink and making an asymmetric set of dots on the circumference of the circle, poking holes through the surface of the filter and into the agarose. This will simultaneously label the plate and the filter and will orient the filter relative to the plaques on the plate. The nitrocellulose paper will shrink upon heating. For reasons that will become obvious, the filters should be preshrunk before they are placed on the phage plates. Do this by boiling them in H_2O for 5 min and blotting to damp dryness or by autoclaving for 5 min. Leave the paper on the agarose for 2 min and then lift it with tweezers. For a second paper replica of the plate, apply another paper for 3 min; for a third, use 4 min. As with Southern transfers, handle the nitrocellulose paper with tweezers and wear gloves so that you can grab the paper should an emergency arise.

6. At room temperature, immerse the replica filters in 0.1 N NaOH, 1.5 M NaCl for 2 min, then 0.2 M Tris-HCl, pH 7.5, 2× SSC for 2 min, and then twice for 15 min in 2× SSC. Blot dry with Whatman 3-MM paper. Bake and, if the filter-bound DNA is to be hybridized to radioactive DNA, soak in prehybridization solution as described in the Southern transfer procedure (p. 154).
7. Hybridization is done at the aqueous (nonformamide) conditions described for Southern transfers. The washing is also essentially the same as for Southern transfer filters. Wash the filters twice in 2× SSC for a few minutes at room temperature. Then wash for 3-6 h at 65 °C in 0.5× SSC, 0.5% SDS. Dry the filters thoroughly and mark the orientation spots with a radioactive compound. Autoradiography is done as described for Southern transfers.
8. Filter paper replicas may be reused by repeating steps 6 and 7; however, in our experience the paper becomes more brittle and frequently cracks. Also, the hybridization signals diminish with successive uses probably because DNA falls off the filters.
9. To be sure that a radioactive spot is the result of specific hybridization, hybridize and autoradiograph two filter transfers from each phage plate. If both autoradiograms give a signal in the same spot, then specific hybridization has usually occurred.

PHAGE RECOVERY PROCEDURE

1. Place the developed x-ray film on a light box and align the phage plate on top of it. Remove the positive agar region by making a core of the plate surface with the mouth of a sterile Pasteur pipette. Dissolve the core in 1 ml of lambda dilution buffer (10 mM Tris-HCl, pH 7.5, 10 mM $MgCl_2$).

2. Titer the phage concentration and then spread at approximately 200 PFU/plate and rescreen paper replicas in order to purify the positive phage. Two additional cycles of picking and screening are advisable when the original screen was done at high phage density. When picking phage plaques from these low-density plates, use the tip of a sterile Pasteur pipette. Always pick and screen one cycle after an easily identifiable and well-separated positive plaque has been found.

GROWING PHAGE STOCKS

Stocks of the desired recombinant phage can be made either on plates or in liquid. The DNA can be purified by the methods described in Chapter 2. In addition to the methods described previously, the following modification of the SDS-DNA extraction method is useful for a quick and dirty extraction of DNA that will be clean enough for restriction enzyme digests.

1. Begin with 10 ml of a plate or liquid lysate with phage at a concentration greater than 10^{10} particles/ml.
2. Add 2 ml of 0.25 M EDTA, 0.5 M Tris-HCl, pH 9.0, 2.5% SDS and incubate at 65 °C for 30 min.
3. Add 0.5 ml of 8 M K^+-acetate and chill on ice for 15 min. Centrifuge at 17,000 rpm in the SS34 Sorvall rotor for 15 min. Decant the supernatant into a clean tube. If any precipitate is carried over into the clean tube, repeat the centrifugation and decant the supernatant. Add 5.6 ml of 95% ethanol, mix and chill at −70 °C for 20 min or at −20 °C for 30 min. Centrifuge at 17,000 rpm in a Sorvall SS34 rotor for 15 min, decant the supernatant and remove any excess liquid with a Q-tip.
4. Dissolve the precipitate in 0.4 ml of 2 M NH_4^+-acetate, using a Pasteur pipette to break up the pellet. Transfer to a 1.5-ml microfuge tube and add 0.8 ml of 95% ethanol. Vortex and precipitate the DNA by the standard method for rapid precipitation of DNA (p. 95).
5. This DNA may be digested with restriction enzymes by using about 1 μl of the DNA solution in 10-μl of reaction volume, adding sufficient restriction enzyme to digest 2 μg of DNA in 30 min, and incubating for 30 min. The DNA product is occasionally contaminated by large amounts of RNA. If this RNA obscures the restriction pattern on gels, then follow the restriction digestion with a digestion by 1 μg of RNase for 15 min at 37 °C.

Screening Colonies

This method (Gergen et al., 1979) is considerably simpler than the original Grunstein and Hogness (1975) method. It is also less expensive and produces filters that can be reused many times. The two methods appear to have about the same maximum sensitivity; however, to achieve this maximum with the method we describe, some attention must be paid to the rate of the hybridization reaction. For most applications, rates can be largely ignored.

PROCEDURE

1. Grow colonies to a diameter of 1–4 mm. Transfer them to Whatman 541 filter paper by placing the paper over the colonies and incubating at 37 °C for 2 h. When placing the paper over the colonies, do not trap air between the paper and the agar surface because an air bubble will prevent transfer.
2. If the plasmids have relaxed control of DNA replication, then amplify the plasmid copy number by transferring the paper, face down, onto LB agar, 250 μg/ml chloramphenicol, and incubating for 24 h at 37 °C. See p. 105 for possible problems with some lots of chloramphenicol.
3. Lyse the colonies and denature and immobilize their DNA on the paper by washing twice with agitation for 5 min each in the following 3 solutions: 0.5 M NaOH, 0.5 M Tris-HCl, pH 7.4; and 2× SSC, pH 7. Wash briefly in 95% ethanol and dry in air. The filters have very high wet strength and the DNA is bound to the filters very rapidly. For these reasons all filters can be treated as a single batch and no special care is necessary.
4. Label the filters with black ink "Sharpie" pens (Sanford's).
5. Hybridize in 50% formamide, 5× SSC, pH 7.5, 250 μg/ml carrier tRNA or DNA in a volume of 75 $\mu l/cm^2$ of filter. In general about 10^5 cpm of [^{32}P]-labelled RNA or DNA/ml of hybridization solution are hybridized with gentle agitation at 37 °C for approximately 24 h. The optimum length of hybridization time will vary with the complexity of the radiolabelled nucleic acid. With these filters it is best to minimize both the concentration of radiolabelled nucleic acid in the hybridization solution and the length of hybridization time, since the amount of nucleic acid that binds tightly and randomly to the 541 filter paper increases with time. Details of the kinetics of this hybridization reaction and of the random binding are given in Gergen et al. (1979). This paper should be referred to if maximum sensitivity is required.
6. Following hybridization, wash the filters four times in a large volume of 2× SSC at room temperature. Filters hybridized to radiolabelled RNA should be digested with 10 μg/ml RNase A from bovine pancreas (Calbiochem) in 2× SSC for 30 mins at 25 °C. Significantly higher digestion temperatures, such as 37 °C, will remove radiolabel from terminally labelled, hybridized RNA. If the filters are to be reused, residual RNase activity can be eliminated by incubation in 0.05% (vol/vol) diethylpyrocarbonate (Sigma) for 30 min at 25 °C.

7. Autoradiograph the washed air-dried filters using the same methods described for DNA gels (p. 184). If the colony positions are not obvious on the autoradiogram, determine them by aligning the autoradiogram with the filters that have been stained with 1 μg/ml ethidium bromide and illuminated with UV light.
8. After use the filters may be recycled by passing them through the entire washing procedure described in step 3. This will remove more than 95% of the hybridized DNA and all of the hybridized RNA. Filters may be reused more than 5 times without a noticeable change in screening sensitivity.

ISOLATING DNA FROM A SINGLE COLONY

It is frequently necessary to screen 10–100 colonies in order to identify one that has a plasmid of a particular size or a particular set of restriction sites. For example, a particular EcoR I restriction fragment may have been cloned into a plasmid EcoR I site and a clone with one of the two possible insert orientations may be required. It would be useful to have a quick method for isolating, restricting, and electrophoresing DNA from 10 different transformed colonies so that colonies containing a plasmid with an insert in the required orientation could be identified. The method described below accomplishes the first step, rapid isolation of restrictable DNA. A single colony will yield DNA that can be seen in an ethidium bromide–stained gel. The method was devised by Holmes and Quigley and modified by Barnett (personal communication). An earlier method is described in the literature (Barnes, 1977).

PROCEDURE

1. Use the end of a toothpick to collect bacteria from a large colony or a streak of bacteria on a plate.
2. Resuspend the cells in 25–50 μl of 50 mM EDTA, 50 mM Tris, pH 8.0, 5% Triton X-100, 8% sucrose. Do this in a 1.5-ml polypropylene microfuge tube. This is a crucial step. Make sure that all of the cells are resuspended.
3. Add an equal volume of the same buffer containing 2 mg/ml of lysozyme. If you are working with only a few samples, it is possible to include the lysozyme in the resuspension buffer. However, do not leave them in this buffer for more than 5 min before you proceed to step 4.
4. Put the tubes in a 100 °C oil bath for 3 min.
5. Cool the tubes on ice for 5 min.
6. Centrifuge for 10–15 min in an Eppendorf microfuge.
7. Remove the supernatant with a mechanical micropipette

(Pipetman or Eppendorf), being careful to avoid the pellet. The pellet size and texture are variable. Deliver the supernatant to a new microfuge tube.

8. Add an equal volume of 2-propanol to the supernatant fraction and place at −20 °C (*not* −80 °C) for 30 min.
9. Centrifuge for 15 min in an Eppendorf microfuge.
10. Decant the alcohol and drain dry. Do not vacuum desiccate or use any other method to completely dry the pellet.
11. Resuspend the DNA in 25 μl of the desired restriction enzyme buffer. There is usually enough DNA to be seen in two different lanes of a gel.
12. After digestion with restriction enzyme(s) add RNase A to 20 μg/ml and incubate for 2 min at room temperature before loading the gel. This last step is not always necessary. It removes a large amount of RNA from the low molecular weight (50–200 nucleotides) region of the gel.

SOUTHERN TRANSFERS

A procedure known as Southern transfer (Southern, 1975) is of great utility in recombinant DNA technology, since it allows physical mapping on DNA of regions complementary to a radiolabelled nucleic acid. The technique accomplishes this by utilizing both the ability of gels to separate DNA according to size and the high specificity of DNA-DNA and RNA-DNA hybridization reactions. In a typical Southern transfer reaction a restriction digest of DNA is separated according to size on agarose or acrylamide gels. After electrophoresis the DNA is denatured, transferred to filter paper, and immobilized. The transfer is carried out in a way that preserves the gel separation pattern on the filter replica. Following transfer, DNA in the filter replica is hybridized to a ^{32}P-labelled nucleic acid and subsequent autoradiography (p. 184) identifies the restriction fragments with complementary sequences. A modification (Thomas, 1980) allows RNA to be detected in agarose gels (Rave et al., 1979).

PROCEDURE FOR TRANSFER

1. Electrophorese, stain, and photograph the gel by the methods described in the previous chapter. Then use a large spatula (see p. 121) to transfer the gel to a glass or plastic dish (4.5 × 22 × 33 cm) with short walls that will allow later removal of the gel without damaging the gel. Either a Pyrex baking dish or Saran Wrap on a Styrofoam grocery tray works well. Soak the gel twice for 15 min in 0.5 M NaOH, 1.5 M NaCl in order to denature the DNA. Then neutralize the gel by soaking twice for 15 min in 3 M NaCl, 0.5 M Tris-HCl, pH 7.0. In the procedure that follows it is necessary to keep grease off the gel and the

nitrocellulose paper. Always use plastic gloves when touching the gel and the nitrocellulose paper. Try to avoid touching the nitrocellulose paper even when wearing the plastic gloves. Broad-bladed tweezers (Millipore) should be used for all manipulations of the nitrocellulose paper. However, wear plastic gloves when manipulating the nitrocellulose paper, so that you can deal with emergencies.

2. Stack plate glass approximately 1 cm deep in the bottom of an 18 × 30 cm Pyrex dish. The glass surface should be somewhat larger than the gel in each dimension. Place 3 wicks of Whatman 3-MM paper (26 × 16 cm) over the glass and into the Pyrex dish as shown in Figure 6.2. On top of the wicks put 10 sheets of 3 MM that are the same width and length as the gel. Add 10× SSC to a level that is just below the top surface of the 3-MM paper stack (approximately 750 ml), and using plastic gloves, press the air bubbles out of the stack of paper. Put the gel on the stack.

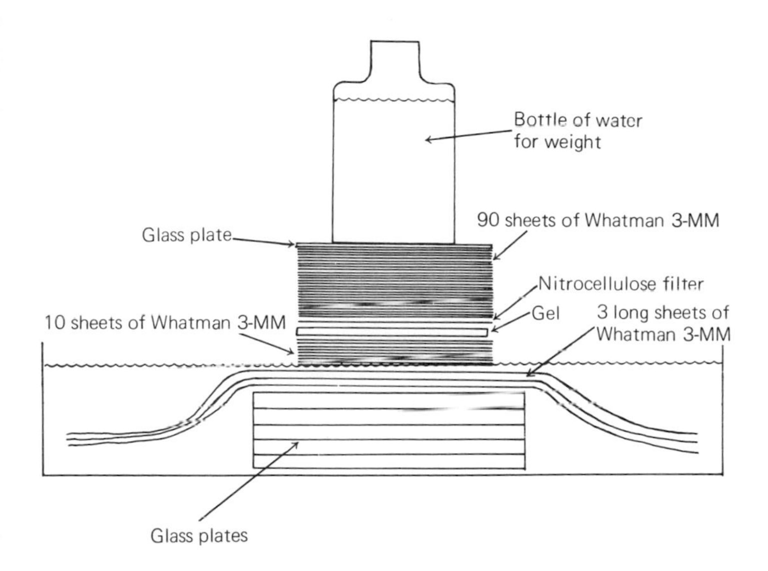

Figure 6.2. The arrangement for transfer of denatured DNA fragments from gel to nitrocellulose filter paper.

3. Cut the nitrocellulose paper to the dimensions of the gel surface. Both Millipore HA and Schleicher & Schuell (S & S) BA85 are fine, but the S & S paper is heavier and less likely to tear and crack. When there is a choice, it is always best to handle and cut the paper when it is damp, as then it is much less brittle.
4. Gently float the nitrocellulose paper on the surface of 2× SSC and allow it to wet. It will turn from white to gray as it wets. If any areas do not wet, rub them lightly with tweezers while the paper is still floating. The usual problem is that an air bubble is trapped under the paper. It will be moved by rubbing. If the entire paper can not be wet, dry it on Whatman 3-MM and then try floating it on 2× SSC again.

5. The objective of this step is to pass liquid through the gel and nitrocellulose filter paper in a way that elutes DNA from the gel to the paper. After the nictrocellulose paper is wet, immerse it in 2× SSC for at least 2 min. Then wet the upper surface of the gel with 2× SSC, place the soaked nitrocellulose paper on the gel, and use tweezers to squeeze out all the air bubbles that are trapped between the gel and the filter. Put a stack of approximately 90 sheets of 3-MM paper (some workers prefer to use 2 sheets of 3-MM and lots of paper towels) on top of the gel. Put a glass plate on top of the stack and weigh it down with approximately 1 kg (a bottle with 1 liter of liquid is usually the most handy weight). The stack of paper below and above the gel should be the same width and length as the gel so that the paper can pass liquid through the entire gel without diversion through a 3-MM-paper short circuit. Let the transfer continue for 12–18 h. The rate of DNA transfer from the gel is analogous to the rate of DNA migration during electrophoresis, that is, the transfer rate is related to the log of DNA length. Most of the DNA with a length less than 1 kb transfers within a few hours and most of the DNA with a length greater than 10 kb is not transferred in 12 h. In most cases the DNA in a gel will have been stained with ethidium bromide and photographed under UV illumination before transfer to filters. This procedure breaks the DNA and speeds the transfer of DNA from the higher molecular weight bands. DNA of high molecular weight can be transferred even more efficiently by partially hydrolyzing it (Wahl et al., 1979) before transfer. The hydrolysis reaction must be limited, however, because short DNA ($<$ 300 bases) does not efficiently bind to the paper.
6. After DNA transfer, remove the nitrocellulose paper with tweezers and soak it in 2× SSC for 15 min. Soaking in 20× SSC will allow somewhat better retention of very short DNA by the filter, but it also makes the salt concentration during subsequent hybridization an unknown. After the soaking, place the filter between 2 layers of 3-MM paper and vacuum dry in an oven at 80 °C for 2 h. Higher temperatures can discolor the paper and will make it brittle. The object of this operation is to thoroughly dry the paper and by some mysterious process thereby bind the DNA to the paper. Overnight incubation at 65 °C without vacuum is also adequate. The transfer paper may be stored dry for months without any loss of usefulness.
7. If the DNA on the paper is to be hybridized to radiolabelled DNA, then soak the dried transfer for at least 6 h at 65 °C in prehybridization solution (0.02% [wt/vol] each of Ficoll 400, polyvinyl pyrrolidone, and BSA [Denhardt, 1966] in a solution of 2× SSC, 50 μg/ml denatured carrier DNA, and 0.5% [wt/vol] SDS). After this soaking, the filter may be used immediately or dried and stored at room temperature. This procedure will reduce the amount of

radiolabelled DNA that binds at random sites on the filter. Its effectiveness varies with different batches of nitrocellulose filter paper, so it is useful to buy moderately large quantities of paper from a batch that works well.

8. Reuse the 3-MM paper for later transfers. Soak it with several changes of a large volume of distilled water, air dry it in stacks of about 20 sheets, and then use a steam iron to flatten stacks of about 5 sheets. If the paper is reused, put at least 5 unused sheets of 3-MM below the gel and above the nitrocellulose paper. Used paper is somewhat corrugated and if put in the bottom of the stack will not give uniform flow of solution through all parts of the gel. If you use paper towels, do not reuse them.

PROCEDURE FOR HYBRIDIZATION

The hybridization reaction is conveniently done in the commercial heat-sealable plastic bags that are normally used for food storage. Sears and Roebuck markets the bags and a heat sealer under the name of Seal-N-Save. Cut the bag so that only one side is sealed and place the nitrocellulose filter onto one inner face of the bag. Seal 2 edges and, for a 17 × 12 cm filter, add 15 ml of hybridization solution, including the radiolabelled probe. Then seal the fourth side of the bag so that as few bubbles and as little air as possible remain in the bag. Three sealed ends of the bag should be less than 1 cm from the edges of the filter and the fourth should be 3 or 4 cm from the edge, as shown in Figure 6.3. Place the bag in a water bath and incubate for the hybridization reaction.

It is important not to have air bubbles on the surface of the filter. Radiolabelled nucleic acid will stick to the paper at the edges of a bubble and will not have a chance to hybridize to filter-bound DNA that is covered by a bubble. For these reasons, use weights to hold one end of the bag at the bottom of the water bath and push all bubbles to the opposite end. The opposite end must have enough room to hold the bubbles, so it should be the one that is 3–4 cm from the edge of the filter. This end will tend to float and the bubbles can be encouraged to rise by pushing them with a blunt object.

We usually hybridize [^{32}P]-RNA to filters in 50% formam-

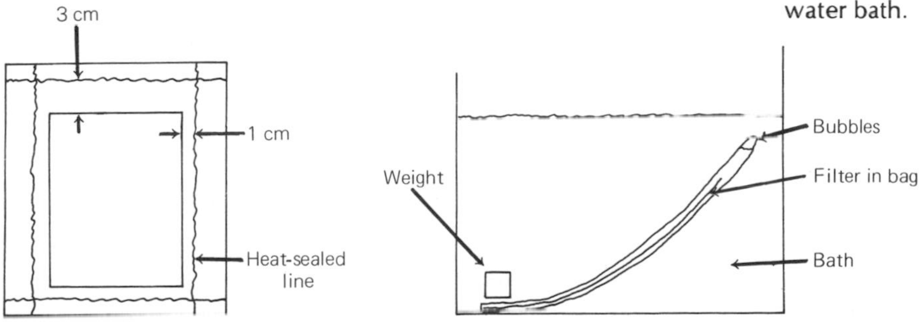

Figure 6.3. The sealing of filter paper in a food storage bag for the hybridization reaction and placement of the bag containing the filter paper in a water bath.

ide, 4× SSC at 37 °C for 18–48 h and [^{32}P]-DNA to filters for the same length of time but at 65 °C and in the solution used to presoak the filter. This gives satisfactorily low backgrounds and high signals. Satisfactory results can be obtained over a wide range of radiolabel concentrations. As a starting point try 10^3–10^4 cpm/ml when using a phage DNA at 10^7–10^8 cpm/μg. These hybridization conditions are nominal and should be adjusted for particular reactions using the guidelines provided above (p. 145). Remember to denature the labelled DNA before hybridization. A 10-min incubation at 100 °C in 0.1× SSC is more than adequate for denaturation.

After hybridization, a filter hybridized in 50% formamide should be washed once in 200 ml 0.5× SSC at room temperature to remove most of the formamide. If the filter has been hybridized to an RNA, incubate it at 25 °C in 2× SSC containing 10 μg/ml RNase A (previously heat treated for 30 min at 80 °C in 2× SSC) for 1 h to digest unhybridized RNA. As mentioned in the colony screening section, elevated digestion temperatures will remove terminally labelled hybridized RNA. RNase-treated RNA-hybridized filters or DNA-hybridized filters are then washed 3 times in 200 ml 0.5× SSC 65 °C for 15 min with gentle shaking. This treatment will usually remove nonspecifically bound radiolabel. If some activity still persists, try several washes for 15 min at room temperature in 3 mM Tris base, pH 9.

The radiolabelled nucleic acid probe can be reused. Save it and denature by boiling for 10 min before reusing.

COMMENTS

1. Different batches of nitrocellulose paper vary considerably in the amount of [^{32}P]-labelled nucleic acid that they will nonspecifically bind. The nonspecific binding usually produces either a general gray background on the autoradiogram or, more frequently, a set of dark spots.
2. This method is extremely sensitive. DNA transferred from a gel band containing 10 pg of a 1-kb DNA fragment will give a good autoradiographic signal (p. 184) with less than a week of exposure (with fluorescent screen at −70 °C) when it has been hybridized under the standard conditions with 10^6 cpm/ml of a pure complementary DNA.

SELECTING RNA COMPLEMENTARY TO A DNA

In some experiments it is necessary to select from a mixture of RNAs an RNA that hybridizes to a particular (cloned) DNA. For example, it is usually possible to positively identify a gene in a cloned DNA segment by characterizing the polypeptide translated in vitro from the mRNA which hybridized to the cloned DNA. This mRNA selection method is also a path to various characterizations of in vivo transcripts of a cloned gene. The most convenient method of selection is to immobilize the DNA on a filter and then to hybridize the total RNA sample to the immobilized DNA. After the filters are washed to remove unbound RNA, that RNA which hybridized to the immobilized DNA can be eluted and translated.

The following solid phase hybridization selection method (Barnett et al., 1980) is based on the diazobenzyloxymethyl (DBM) paper immobilization of single-stranded nucleic acid (Alwine et al., 1977), a method most frequently used for Northern gels. Denatured DNA immobilized on a piece of DBM paper can be used for many successive hybridization selections of RNA. We have not quantitated the technique but, in terms of one practical application, 0.4 μg of complementary DNA bound to DBM paper selected a specific RNA that could be translated into a polypeptide that gave a strong autoradiographic signal after a 10-h fluorographic exposure.

The method is described for a 1-cm^2 piece of paper. It may be scaled up by multiplying all sizes and volumes by the same factor.

Preparing DNA

Dissolve 2–100 μg of DNA in a 1.5-ml microfuge tube with 12.5 μl of distilled water. Add 1 μl of 4 N NaOH, close the tube cap, and heat at 100 °C for 5 min. Put the tube in an ice water bath and add 10 μl of 100 mM Na^+-phosphate, pH 6 and 4 μl of 1 N HCl. Adjust the volume to 25 μl with distilled water and freeze until ready for use.

Preparing NBM Paper

Cut a 1-cm^2 piece of Whatman 540 or 541 paper and place it in a small Pyrex baking dish which is placed in a 60 °C water bath. Add 2.3 mg of *m*-nitrobenzylpyridinium chloride (NBPC) and 0.7 mg Na^+-acetate trihydrate to 28.5 μl of distilled water in a microfuge tube. Pipette this solution onto the paper and push bubbles from under and over the paper by rubbing with gloved hands. Do not touch the solution with your skin. Continue to rub the paper until it is dry. This will allow the NBPC to be evenly distributed over the paper. Drying will take about 5 min. Hang the paper in a 60 °C oven for 10 min. Remove the paper from the oven, readjust the oven to 130–135 °C, and then bake the paper for 30 min. Wash the paper at room temperature for 10 min in 10 ml of distilled water 2 times and for 10 min in 10 ml of acetone 3 times. Allow the paper to dry in air. This NBM paper is stable for many months if it is stored at 4 °C.

Converting NBM Paper to DBM Paper and Binding the DNA

PROCEDURE

1. Incubate 1 cm^2 of NBM paper at 60–65 °C for 30 min with occasional shaking in 0.4 ml of 20% (wt/vol) sodium dithionite (sodium hydrosulfite). Perform this incubation in a hood so that SO_2 will not accumulate.
2. Decant the dithionite and wash the paper 4 times for a few minutes in about 20 ml of distilled water. Then wash in 20 ml of 1.2 M HCl at 4 °C.
3. Soak the paper for 30 min with occasional shaking in Na^+-nitrite solution at 4 °C. The solution should be made immediately before use by adding 2.7 ml of 10 mg/ml $NaNO_2$ in water to 100 ml of 1.2 M HCl.
4. Decant the nitrite solution. Wash the filter in 40 ml of distilled water and then in the same volume of 20 mM Na^+-phosphate, pH 6. The product is called DBM paper.
5. Working rapidly, blot the filter to damp dryness with a paper towel and place it in a siliconized scintillation vial or some other siliconized glass container. Immediately add the thawed DNA solution to the filter, cap the vial, and leave it at room temperature overnight. This may be an excessive time period because it is likely that the binding is accomplished in only 1–2 h.
6. Wash the 1-cm^2 paper in about 50 ml of distilled water for a few minutes and then incubate it in 4 successive 10-min washes of 10 ml 0.4 N NaOH at 37 °C.
7. Wash in distilled water to remove the base. Store the paper at 4 °C in hybridization buffer (50% formamide, 4× SET [SET is 150 mM NaCl, 1 mM EDTA, 10 mM Tris-HCl, pH 7.8, 0.1% SDS]).
8. Sometime before the hybridization reaction incubate the 1-cm^2 paper in 5 ml hybridization buffer at 37 °C for 2–5 h. Decant the hybridization buffer, add 5 ml of 99% formamide, and incubate at 65 °C for 30 min. Filters may be stored at 4 °C or used immediately for hybridization.

Hybridizing RNA to DBM-Bound DNA

The considerations described on p. 145 (General Aspects of Nucleic Acid Hybridization Reactions) apply to this reaction. In general, a useful procedure is to hybridize a 1-cm^2 paper at 37 °C for 12–24 h in 75 μl of hybridization buffer containing 500 μg/ml of poly-A^+ RNA or 5 mg/ml of total RNA. These conditions work well for hybridization selection of moderately abundant mRNAs from *Drosophila melanogaster.* Place the filters on Parafilm, add hybridization solution, and then cover the filters with Parafilm and incubate at 37

°C. Larger volumes and more filters require siliconized vials that are gently shaken during the reaction.

Recovering the RNA

PROCEDURE

1. After hybridization, save the reusable RNA hybridization solution.
2. Wash the 1-cm^2 paper in 10 ml of 2× SET at room temperature, then incubate with shaking in 5 ml of 50% formamide, 0.2× SET, 0.1% SDS at 37 °C for 30 min. Repeat this incubation 3 times. Note that this is a fairly stringent washing criterion, that is, it is fairly close to the T_m of most well-paired duplexes.
3. Remove the filter and drain by touching it to a paper tissue. Put the filter in 0.2–1 ml of 99% formamide, 10 mM Tris-HCl, "pH 7.8," that has been made by adding Tris-HCl, pH 7.8, from a 2 M stock solution. Incubate at 65 °C for 5 min.
4. Remove the solution, place it in a 15-ml siliconized Corex tube, and add to it the following: (a) an equal volume of sterile distilled water; (b) Na^+-acetate, pH 5.0, to make 0.3 M; (c) carrier tRNA to 10 μg/ml; and (d) 2.5 vols of 95% ethanol. Then store overnight at −20 °C. Precipitate the RNA at 10,000 rpm in an HB4 or SS34 Sorvall rotor. Wash once with 70% ethanol, drain, dry under vacuum, and redissolve the RNA in distilled water in preparation for translation.

Synthesis of NBPC (nitrobenzylpyridinium chloride)

This is a very expensive compound that can be synthesized easily. The synthesis is described by Alwine et al. (1977). There are some helpful hints in the method given below. Do the entire procedure in a well-vented chemical hood.

PROCEDURE

1. To 1 liter of benzene add 158 g of paraformaldehyde and 200 g of *m*-nitrobenzyl alcohol.
2. Bubble dry HCl gas (454 g) through the above solution for 2–3 h at room temperature with stirring, as shown in Figure 6.4. HCl gas is generated from the reaction of H_2SO_4 with NaCl.
3. Continue stirring overnight.
4. Stop mixing and allow the phases to separate (10–15 min).

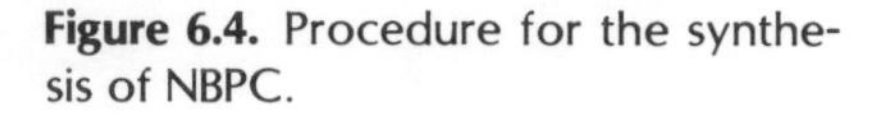

Figure 6.4. Procedure for the synthesis of NBPC.

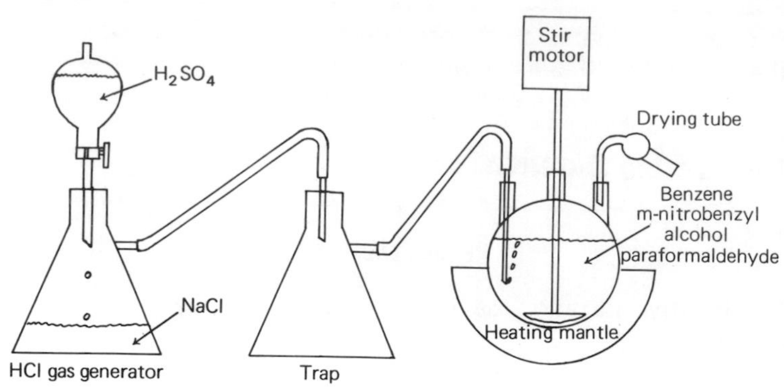

5. Remove the upper organic phase using a separatory funnel.
6. Pour the organic phase into a 2000-ml flask and dry it by adding 150 g of Na_2SO_4 (anhydrous) with stirring.
7. Gravity filter the suspension into a 1000-ml round-bottom flask.
8. Remove benzene under reduced pressure in a rotary evaporator.
9. Distill the remaining yellow liquid as shown in Figure 6.5 under reduced pressure. At a pressure of 1.5 mm Hg collect the fraction distilling between 150 and 154 °C.
10. Add the yellow distillate slowly to 750 ml of ice cold pyridine with stirring and allow pyridine salt to crystallize overnight.
11. Collect the salt on a coarse sintered glass filter.

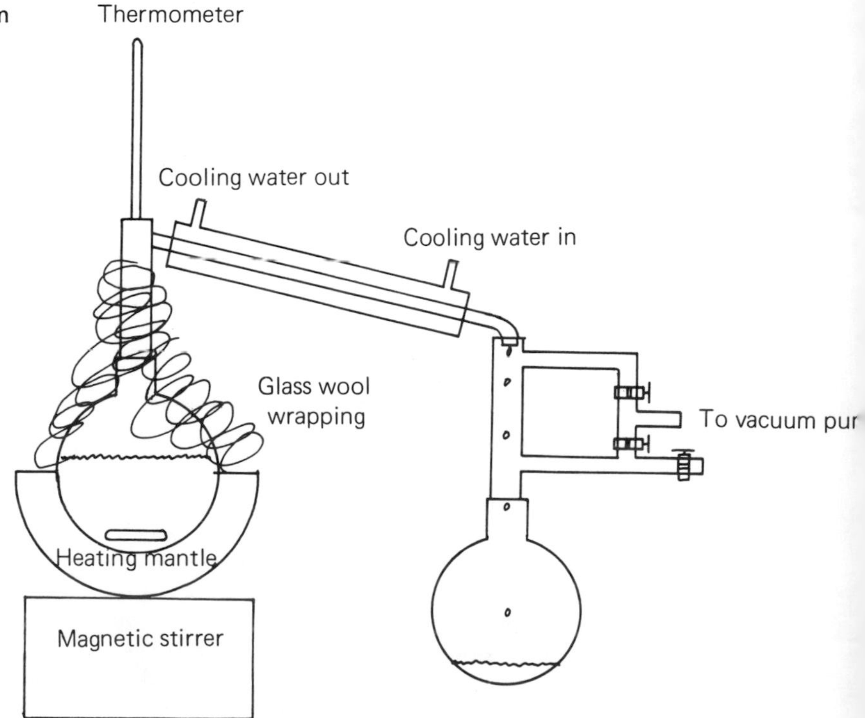

Figure 6.5. The procedure for vacuum distillation purification of NBPC.

12. Wash first with pyridine and then several times with 2 quarts of petroleum ether.
13. Dry the product overnight under reduced pressure in a desiccator.
14. The yield should be about 200 g.

IN VITRO TRANSLATION SYSTEMS FROM HIGHER ORGANISMS

Several systems from higher organisms have been developed for translating RNA. We will describe the wheat germ and rabbit reticulocyte systems. Both work well and can be used routinely. The reticulocyte lysate is more difficult to prepare and more expensive. Rabbits are the costly component. In general, the reticulocyte system is more efficient in translating messages and this is particularly true for messages that code for proteins larger than 70,000 daltons. The reticulocyte system also is less likely to terminate translation prematurely. If one or a few mRNAs are to be translated frequently it is worth the effort to prepare both lysates and test which is more efficient in translating those particular mRNAs.

Preparing a Wheat Germ Extract

This preparation method is modified from Roberts and Paterson (1973).

RECIPES AND MATERIALS

1. *Untoasted wheat germ* from W. C. Mailnot, General Mills, Inc., Wayzata Blvd., Minneapolis, Minn. Different batches vary in activity. Store the wheat germ under vacuum at 4 °C because moisture adsorption by the germ will lead to reduced translation activity of the lysate.
2. *Grinding buffer*
   20 mM Hepes (Calbiochem), pH 7.6, adjusted with KOH
   1 mM Mg^{2+}-acetate
   100 mM KCl
   2 mM $CaCl_2$
   6 mM 2-mercaptoethanol
3. *Column buffer*
   20 mM Hepes, pH 7.6, adjusted with KOH
   120 mM KCl
   5 mM Mg^{2+}-acetate
   6 mM 2-mercaptoethanol
4. *Master translation mix* (wheat germ). Store in aliquots at −70 °C.
   100 μl 10 mM all amino acids except methionine

200 μl 200 mM ATP, neutralized with KOH
10 μl 100 mM GTP, 0.01 M Hepes, pH 7.2, adjusted with KOH
200 μl 1.6 M creatine phosphate, 0.01 M Hepes, pH 7.2, adjusted with KOH
80 μl 1 M dithiothreitol
960 μl 1 M Hepes, pH 7.2, adjusted with KOH
50 μl 200 mM spermidine (free base)
1600 μl distilled H_2O
3.2 ml total volume

5. *Creatine phosphokinase* (Sigma), 5.4 mg/ml in 50% glycerol.

PROCEDURE

Use sterile glassware and buffers. Note that many buffers should not be autoclaved and must be sterilized by filtration. Millipore filters that have been boiled two times in distilled water (to remove plasticizers) are adequate. In general, the best lysates result from rapid preparation at as low a temperature as possible.

1. Pre-chill a mortar and pestle to 2 °C and perform the following in a cold room (not warmer than 4 °C) in a bucket of ice. Add 12 g of wheat germ and 12 g of sea sand to the mortar. The sea sand should be baked overnight at greater than 100 °C before it is used. Begin grinding the wheat germ and then, while continuing to grind, slowly add 28 ml of grinding buffer (2 °C) in 5-ml aliquots over a period of 15 min. This will result in a thick paste.
2. Centrifuge the paste in 30-ml Corex tubes, 2 °C at 30,000 × *g*, in the SS34 Sorvall rotor for 10 min. This is beyond the manufacturer's advertised force limit for Corex tubes, so use new tubes and in the future only use them for wheat germ preparations. Do not let them get scratched, wash them by hand, be gentle, and they will last forever.
3. Remove the supernatant avoiding both the pellet and the surface layer of fat. To this volume of approximately 10 ml, add creatine phosphokinase to 40 μg/ml. Incubate at 30 °C for 15 min.
4. Load this extract onto a Sephadex G-25 (medium) column (3.5 × 35 cm) that is preequilibrated with column buffer at 4 °C. The column is run at 4 °C in a cold room with a flow rate of 1 ml/min and 1-ml fractions are collected.
5. Pool the most turbid fractions and freeze in small droplets by dispensing the extract through a sterile syringe into liquid nitrogen. The extract can be stored at −70 °C for several years without substantial loss of activity.
6. If control incubations indicate that the extract has bothersome levels of endogenous mRNA, a micrococcal nuclease digestion, such as that used in the reticulocyte extract preparation, should be used. Adjust the extract to

1 mM $CaCl_2$, add 50 units of nuclease/ml of extract, and incubate for 10 min at 20 °C. Add EGTA to 2 mM and freeze as described above.

Wheat Germ Translation Reaction

PROCEDURE

1. The reaction mixture is made in a 1.5-ml polypropylene microfuge tube sitting on ice. Add 0.1–1 μg of RNA to the tube. If the RNA is in a small volume of distilled H_2O, complete the reaction mixture by adding the components listed below. If the volume of water is too large, remove it by vacuum evaporation. If the RNA must be precipitated from a buffer and salt solution, wash the precipitated RNA with 70% ethanol in order to remove most of the salt. Then add the following:
   2 μl master translation mix
   0.5 μl 25 mM Mg^{2+}-acetate
   2 μl 1 M K^+-acetate
   2–5 μl [^{35}S]methionine (Amersham, 5–6 mCi/ml, 950–1300 Ci/mmol)
   distilled water to bring to a final volume of 17 μl
2. Incubate for 5 min at 22 °C, add 8 μl of lysate, and incubate for 1.5–3 h. Stop the reaction by placing it on ice. Freeze it for longer term storage.
3. Assay the incorporation of label by spotting 1 μl onto one-half of a 2.4-cm diameter Whatman 3-MM filter circle, soaking the filter for 1 min in 10% TCA, 4 °C, boiling in a large volume (approximately 50 ml/filter) of 5% TCA for 10 min and then rinsing it in ethanol and finally in ether before drying and counting in a toluene-based scintillation fluor. Many filters can be washed simultaneously in these solutions.
4. Endogenous activity usually gives incorporation of about 5000 cpm/μl. A good lysate will incorporate 6–10 times more radiolabelled precursor when mRNA is included in the reaction. If 100,000 cpm of the translation product is electrophoresed in a polyacrylamide gel and fluorographed, the translation products should be detectable in 1 day.

Preparing a Rabbit Reticulocyte Lysate

This procedure is based on a method described by Pelham and Jackson (1976).

REAGENTS

1. *Salt stock*
   0.14 M NaCl

1.5 mM Mg^{2+}-acetate
5 mM KCl

2. *Acetylphenylhydrazine* (Sigma)
   1.2% freshly dissolved in salt stock and neutralized with 1 M Hepes, pH 7.5
3. *Heparin* 25 units/ml
   made by dilution of commercial solution, 10,000 units/ml (Ely Lilly), into salt stock

PROCEDURE

1. Make 3–4 New Zealand White rabbits (4–6 lb each) anemic by subcutaneous injection with 1.2% acetylphenylhydrazine according to the following schedule: 2 ml per rabbit on the first day; 1.6 ml on the second; 1.2 ml on the third; 1.6 ml on the fourth; and 2 ml on the fifth.
2. Bleed each rabbit on the seventh, eighth, and ninth day. To do this, first shave one ear and swab it with xylene-saturated cotton. Then use a new razor blade to make a single incision to the posterior ear vein midway along the length of the ear. Collect the blood from each rabbit into 50 ml of ice cold 25 units/ml heparin. Each rabbit should yield approximately 50 ml of blood.
3. Filter the blood through cheese cloth and harvest the cells by centrifugation at 2000 × *g* for 5 min at 4 °C. Wash the cells 3 times by resuspending in the same heparin solution and pelleting. The final centrifugation should be at 8000 × *g* to pack the cells tightly.
4. Lyse these packed cells by the addition of an equal volume of water at 2 °C and, after 1 min, centrifuge the lysate at 30,000 × *g* for 20 min at 2 ° C.
5. The supernatant of this lysate may be frozen (0.5-ml aliquots are convenient) at −70 °C or, if possible, digested immediately with nuclease.

Micrococcal Nuclease Digestion of Lysate

This procedure is based on that of Pelham and Jackson (1976).

REAGENTS

1. *Micrococcal nuclease* (Worthington)
   150,000 units/ml in 50 mM glycine, pH 9.0, 2.5 mM $CaCl_2$. Stored at −20 °C.
2. *Creatine phosphokinase, type I* (Sigma)
   40 mg/ml in 50% glycerol
3. *Hemin* 4 mg/ml in ethylene glycol
   Prepare the stock by dissolving 20 mg hemin in 0.4 ml 0.2 M KOH, then add 0.6 ml H_2O and mix, then add 0.1 ml 1 M Tris-HCl, pH 7.8 and adjust to pH 7.8 with HCl if necessary. Finally add 4 ml ethylene glycol.

After these additions centrifuge at 2000 × *g* for 5 min and discard the pellet. Do not store this solution.

PROCEDURE

Make a reaction mixture from 100 parts lysate, 2 parts 50 mM $CaCl_2$, and 1 part 4 mg/ml hemin. Bring the reaction mixture to 20 °C, add 0.05 parts micrococcal nuclease, and incubate for 15 min. Then add 2 parts 100 mM EGTA and 0.4 parts 40 mg/ml creatine phosphokinase. Make aliquots and freeze at −70 °C. The lysate remains active through several freeze-thaw cycles.

The hemin concentration in this reaction that will lead to optimal activity of the lysate varies from rabbit to rabbit. One ml is a nominal value. The first lysate (day 7 bleeding) from each rabbit should be titered for hemin concentration (0.5, 1.0, 1.5 ml/100 ml lysate, for example) that gives the lysate with maximum translation activity. This optimal amount should be used for lysates from the day 8 and day 9 bleedings.

Reticulocyte Translation Reaction

This procedure is derived from that of Paterson et al. (1977).

RECIPES

1. Creatine phosphokinase 10 mg/ml in 50% glycerol. Store at −20 °C.
2. Master mix (reticulocyte). Store in aliquots at −70 °C.
   66 μl 37.5 mM ATP, 15 mM GTP, neutralized with KOH
   15 μl 100 mM spermidine
   50 μl 400 mM creatine phosphate
   50 μl 1.25 M Hepes, pH 7.6, adjusted with KOH
   25 μl 200 mM dithiothreitol
   12.5 μl 5 mM of all amino acids except methionine
   300 μl 1 M K^+-acetate
   135 μl 32.5 mM Mg^{2+}-acetate
   320 μl distilled H_2O

PROCEDURE

1. Mix 1 μl of 10 mg/ml creatine phosphokinase and 39 μl of master mix in a 1.5-ml polypropylene microfuge tube on ice.
2. The reaction mixture is also made in a 1.5-ml microfuge tube on ice. It requires the following:
   0.1–2 μg of mRNA in H_2O or ethanol, precipitated and washed with 70% ethanol and dried
   10 μl of the mixture of creatine phosphokinase and master mix
   10 μl of lysate

2–5 μl of [^{35}S]-methionine (Amersham), 5–6 mCi/ml, 950–1300 Ci/mmol

3. Incubate the reaction at 37 °C for 1 h and measure the incorporation of label as described for the wheat germ translation. A 10-fold stimulation of incorporation by addition of mRNA is very good.

PURIFYING TOTAL AND POLYSOMAL RNA

There are many different procedures for purifying RNA that is suitable for crude hybridization reactions. The procedures differ considerably in their yield of RNA that is intact and suitable for translation reactions or other reactions requiring full-length RNA. Described below are two procedures for purifying intact RNA from *Drosophila.* One of these is for extraction of total RNA from adults and the other is for extraction of polysome-bound RNA from embryos. Although these procedures have proven very reliable with *Drosophila* they may only be useful as a general guide for extracting RNA from other species, because each species has different nuclease concentrations, cell organization, and so on.

Extracting RNA from *Drosophila* Adults

This method is a modification of that described by Barnett et al. (1980) which was based on a procedure of Tartof and Perry (1970).

PROCEDURE

Use only sterile solutions and bake all glassware.

1. Put 3–5 g of live flies into 30 ml of RNA extraction buffer (0.15 M Na^+-acetate, 50 mM Tris–HCl, pH 9.0, 5 mM EDTA, 1% SDS, 20 μg/ml polyvinyl sulfate) at room temperature in an 85-ml Omnimixer cup (Sorvall). Just before homogenization, add diethylpyrocarbonate (DEPC) to a final concentration of 1%.
2. Homogenize the flies at full speed for 3 times in 30-sec bursts. Intersperse the bursts with 30-sec periods of cooling on ice.
3. Add 25 ml of extraction buffer–saturated phenol at 4 °C to the cup and homogenize for an additional 45 sec. *Do all subsequent steps at 4 °C and prechill all solutions to 4 °C.*
4. Decant the homogenate into a centrifuge tube or bottle and shake for 10 min; then centrifuge for 10 min at

10,000 × *g* to separate the phases.

5. Reextract the interphase with an additional 10–15 ml of the extraction buffer (no DEPC) and recentrifuge.
6. Combine the supernatants from both centrifugations, extract with phenol, and centrifuge as above. Repeat this phenol extraction 2–3 times.
7. Extract the resulting aqueous phase with an equal volume of phenol-chloroform-isoamyl alcohol (24:24:1), separate the phases by centrifugation, and decant the supernatant.
8. Add Na^+-acetate to make this supernatant 0.15 M and precipitate the RNA by adding 2.5 vols of 95% ethanol at −20 °C. Store at −20 °C for 2 h and then pellet the RNA by centrifugation at 10,000 × *g* for 10 min.
9. Resuspend the RNA in 10 ml of 10 mM Tris-HCl, pH 7.6, 1 mM EDTA, 0.1% SDS. Pour this solution into DEPC-treated centrifuge tubes (p. 168) and pellet the large polysaccharides by centrifugation in a Beckman SW50.1 rotor at 25,000 rpm for 45 min at 2–4 °C.
10. Decant the clear supernatant, add proteinase K (Merck) to 100 μg/ml, and incubate for 20 min at 37 °C. Following this incubation, extract the solution twice with phenol and once with phenol-chloroform-isoamyl alcohol, as described above. Add Na^+-acetate and precipitate as described above. Wash the RNA pellet once with 70% ethanol and then store in 95% ethanol at −20 °C.

Extracting Polysomal RNA from *Drosophila* Embryos

This method is modified from Lis et al. (1978). Use baked glassware, sterile solutions, and precooled rotors and tubes.

RECIPES

1. Make the polysome extraction buffer by sterilizing a 10-fold concentrated solution of KCl, $MgCl_2$, and Tris-HCl, pH 7.6 and then adding the sucrose and sterile distilled water to achieve the final concentrations listed below.
   0.25 M KCl
   0.025 M $MgCl_2$
   0.05 M Tris-HCl, pH 7.6
   0.25 M sucrose, RNase free (Schwarz/Mann)
2. RNA extraction buffer. After mixing, adjust the pH of this solution to 9.0 with HCl.
   0.15 M Na^+-acetate
   0.05 M Tris-HCl
   0.005 M EDTA
   1% SDS
   20 μg/ml polyvinyl sulfate

PROCEDURE

1. Dechorionate the embryos by rinsing in 0.7% NaCl, 0.01% Triton X-100 and then soaking for 2 min in a 2-fold dilution of Chlorox (5.25% Na^+-hypochlorite) in distilled water. Wash the embryos in the NaCl–Triton solution to remove most of the hypochlorite. The rinsing, soaking, and washing can be conveniently done in the homemade clamp for nylon cloth, shown in Figure 5.6.
2. Resuspend every 10 g of embryos in 25–50 ml of polysome extraction buffer and then break them open by gently homogenizing with 15–20 strokes of a motorized Teflon homogenizer.
3. Filter the homogenate through sterile nylon cloth (mesh 183 μm; Tetko) or through several layers of cheesecloth. Then centrifuge at 10,000 $\times$ *g* for 20 min in polypropylene bottles or Corex tubes.
4. Pour the supernatant into a Beckman SW25.2 centrifuge tube that has been pretreated with 0.2% diethylpyrocarbonate and thoroughly rinsed with sterile H_2O. Underlay the supernatant with approximately 15 ml of sterile 50% (wt/vol) sucrose in polysome extraction buffer.
5. Pellet the polysomes by centrifuging for 5 h at 25,000 rpm and 4 °C in the SW25.2 rotor. Decant the supernatant, quickly wash the polysome pellet with ice cold polysome extraction buffer, and then discard the buffer.
6. Drain the pellet and then suspend it in RNA extraction buffer, using approximately 10 ml of buffer for a pellet derived from 20 g of dechorionated embryos. Shake this tube until the pellet is completely suspended. This often takes several hours at room temperature.
7. Extract the RNA suspension 3 times with equal volumes of buffer-saturated phenol and once with an equal volume of phenol–chloroform–isoamyl alcohol (24 : 24 : 1).
8. Put the supernatant into DEPC-treated centrifuge tubes for the Beckman SW40 rotor and pellet the large polysaccharides by centrifugation at 40,000 rpm for 45 min at 20–25 °C.
9. Add Na^+-acetate to yield a final concentration of 0.15 M and precipitate the RNA by adding 2.5 vols of 95% ethanol and storing at −20 °C for at least 2 h.
10. Pellet the RNA by centrifugation at 10,000 $\times$ *g* for 10 min and then drain and wash the pellet twice with 70% ethanol. Store the RNA in 95% ethanol at −20 °C.

PURIFICATION OF POLY-A$^+$ (mRNA-ENRICHED) RNA

Only a small fraction of the RNA extracted from eukaryotic cells (usually $< 5\%$) is messenger RNA. The vast majority is ribosomal RNA, 5S RNA, tRNA, and various nuclear RNAs. It is possible to separate messenger RNA from this excess of other RNA by using the fact that almost all messenger RNAs contain a poly-A tail on the 3′ end. This tail can be used to separate the messenger RNA by chromatography on an oligo-(dT)-cellulose column. The RNA extract can be passed through a column under conditions that allow the poly-A tails to anneal to the oligo-(dT) on the column, thereby binding the poly-A-containing RNA (poly-A$^+$ RNA) to the column. The unbound RNA can then be washed from the column and finally the bound RNA can be eluted at a salt concentration that is sufficiently low to melt the poly-A-oligo-(dT) duplexes.

The method described is essentially that of Aviv and Leder (1972).

RECIPES

1. *Oligo-(dT) binding buffer* (1×)
   10 mM Tris-HCl, pH7.6
   1 mM EDTA
   0.4 M NaCl
   0.1% SDS
2. *Oligo-(dT) binding buffer* (2×)
   10 mM Tris-HCl, pH 7.6
   1 mM EDTA
   0.8 M NaCl
   0.1% SDS
3. *Oligo-(dT) elution buffer*
   10 mM Tris-HCl, pH 7.6
   1 mM EDTA
   0.1% SDS

PROCEDURE

Unless otherwise specified, all operations are done at room temperature.

1. Thoroughly drain ethanol from the RNA pellet and then resuspend the RNA in oligo-(dT) elution buffer. Measure the absorbance at 260 and 280 nm. The A_{260}/A_{280} should be about 2.
2. Adjust the RNA solution to 5-10 mg/ml with elution buffer and heat for 5 min at 65 °C, quick-cool, and mix with an equal volume of 2× oligo-(dT) binding buffer.
3. Pass the RNA solution through an oligo-(dT)-cellulose column (Collaborative Research; about 2 g in a column of 0.9 cm diameter) which has previously been washed with 5 column volumes of 1× oligo-(dT) binding buffer. The column is run at a rate of about 10-15 ml/h. The method can be scaled down by using as a column an Eppendorf pipette tip with siliconized glass wool at the bottom.

4. Recycle the column effluent once to give any unbound poly-A^+ RNA a second chance to bind to the column.
5. Wash the oligo-(dT) column with about 5-8 column volumes of 1× oligo-(dT) binding buffer, elute the RNA with elution buffer (usually 2-3 column volumes), and collect with the aid of a Gilson fraction collector (about 25-35 drops per fraction; RNA starts to elute between fractions 16 and 25).
6. Measure the absorbances of the fractions and pool the UV-absorbing fractions.
7. Note that at this point one can rechromatograph the pooled RNA fractions on an oligo-(dT)-cellulose column to further purify the poly-A^+ RNA (step 7a); or one can stop, ethanol-precipitate the RNA, and store it until used or further purified on oligo-(dT)-cellulose (step 7b).

7a. Rechromatography: After step 6 determine the volume of the pooled fractions, heat the solution to 65 °C for 5 min, quick cool, and dilute with an equal volume of 2× oligo-(dT) binding buffer. Pass this sample through the same oligo-(dT) column which has been thoroughly washed with several column volumes of elution buffer and several volumes of binding buffer. Then follow with steps 3-6 and finally precipitate the RNA by adding Na^+-acetate to 0.3 M and 2.5 vol. of 95% ethanol.

7b. Ethanol precipitation followed by rechromatography: If after step 6 a stopping point is desired, add Na^+-acetate to 0.3 M and then add 2.5 vol of 95% ethanol. Store at least 2 h at −20 °C and then pellet the RNA by centrifugation at 10,000 × *g* for 15 min. Wash twice with 70% ethanol and drain. Process the RNA starting with step 1 above, but the initial concentration of poly-A^+ RNA should be 1 mg/ml or less.

8. After use, the oligo-(dT) column should be cleaned by washing it with 2-3 column volumes of 0.1 N KOH and then washing with 5-7 column volumes of 1× oligo-(dT) binding buffer containing 0.02% sodium azide. Store the column at room temperature.

RNA SIZE FRACTIONATION BY SUCROSE GRADIENT CENTRIFUGATION

One convenient method for the fractionation of up to 1 mg of RNA is to separate the RNA according to size by centrifugation through a sucrose gradient. The different fractions then yield different size classes of RNA. The sucrose fractionation procedure described below (modified from Haseltine and Baltimore, 1976) appears to separate RNA according to its molecular weight. The DMSO treatment disaggregates the RNA.

Sterilize all solutions either by adding diethylpyrocarbonate to 0.1% and incubating at 35–45 °C for several hours, at which time no odor of DEPC should remain, or by autoclaving a 10-fold concentrated buffer solution (10× buffer: 1.0 M NaCl, 10 mM EDTA, 5 mM Tris-HCl, pH 7.5, 5% SDS) and a stock of distilled water. The distilled water and concentrated buffer can then be mixed and combined with RNase-free sucrose (Schwarz/Mann) to give the desired solutions. Sterilizing buffers containing sucrose will, of course, lead to caramelized solutions.

Making the Sucrose Gradients

The following description is designed for gradients in SW40 (Beckman) tubes.

PROCEDURE

1. Place 6.3 ml of 15% and 5.9 ml of 30% sucrose solutions into the left and right chambers, respectively, of a DEPC-treated plastic gradient maker. The gradient maker is of the same design as that used to make gradient acrylamide gels for protein separations. Be sure to eliminate bubbles which may form in the channel between the two wells. These bubbles usually can be prevented by prefilling the channel with 15% sucrose in buffer.
2. Put a sterile Teflon magnetic stirring flea in the right, that is, the 30% chamber, and use a magnetic stirrer to force it to stir the solution in that chamber.
3. Adjust the plastic tubing outlet from the right chamber so that it touches the lip of a DEPC-treated SW40 tube about 3–5 mm from its top. Open the outlet valve and when the 30% solution begins to flow, open the channel between the two chambers. The flow rate should allow the centrifuge tube to be filled in 15–20 min. If the gradient forms too quickly, excessive mixing of the layers will result, which can lead to a loss of resolution in the gradients.
4. It is useful to let the gradients sit at room temperature for about 1 h before loading a sample for centrifugation. This will remove inhomogeneities in the gradient.

Preparing the RNA

PROCEDURE

1. Pellet the RNA to be fractionated. Pellet it from an ethanol solution in a siliconized tube, drain briefly, use a sterile Q-tip to remove excess ethanol, and then resuspend the pellet before it dries. The resuspension should be in 99% DMSO, 10 mM Tris-HCl, pH 7.5. Up to 100 μg of RNA can be solubilized in 0.2 ml of this DMSO solution. If a

larger amount of RNA is solubilized in this volume, aggregation usually becomes a serious problem. If the RNA does not dissolve at this concentration, it is best to dilute the solution with more buffer.

2. Heat the RNA in the DMSO solution for 5 min at 37 °C. Then dilute 4-fold with 5 mM Tris-HCl, pH 7.5, 1 mM EDTA, 0.5% SDS, and heat at 65–70 °C for 1 min. Then quick-cool in ice before layering onto a gradient.

Sedimenting and Collecting the RNA

PROCEDURE

1. A general guide to centrifugation conditions is given by the sedimentation of 18S and 28S rRNA at 25 °C in these gradients. At 38,000 rpm for 16 h in an SW40 rotor the 28S RNA will pellet and 18S RNA will be within the last one-fifth of the gradient. At 30,000 rpm for 16 h the 28S RNA will be about three-quarters of the way down the gradient and 18S will be in the middle. Centrifuging at 28,000 rpm for 14 h or 40,000 rpm for 6 h will put the 18S RNA about one-third to one-half of the way down the gradient.
2. Collect 0.5-ml fractions and follow the RNA profile by its UV absorbance. The RNA can be recovered by adding Na^{+}-acetate to 0.25 M and precipitating with ethanol as usual. Carrier RNA may have to be added to achieve efficient precipitation of RNA that is at very low concentrations.

Chapter 7

Assorted Laboratory Techniques

In this chapter we describe some of the widely used, but ill-described techniques or bits of information which are helpful or necessary for successfully using the procedures discussed in the first six chapters. Much of this information is well known in laboratories that practice molecular biology or molecular genetics, but since much of this information is transmitted almost entirely by word of mouth, it is presented here so that it can be available to groups first beginning these types of experiments.

GLASS AND PLASTIC CONTAINERS

Although glass and plastic technologies have produced many varieties of material for containers, most commercial labware is limited to a few materials. The variation in glass (Wheeler, 1958) is due to Na_2O and other impurities that are added to SiO_2 in order to lower glass viscosity and thereby ease the manufacture of glass products. Generally, a lower concentration of impurities leads to a harder glass that is more resistant to chemicals and scratching. The hard glass also has a lower thermal coefficient of expansion, so it is less likely to break when shocked by a temperature change. Hard glass with brand names such as Pyrex, Kimax, and Vitro is useful for storing all reagents and solutions except those that are used directly from the manufacturer's container. Manufacturer's containers are usually soft glass and reusing them can lead to disaster. They usually leach salts, even after extensive soaking and washing.

The three plastics most commonly used in place of glass are polyethylene, polypropylene, and polystyrene. The inexpensive polyethylene and polystyrene have low melting temperatures and melt when autoclaved. In addition, many commercial polystyrenes leach significant quantities of oil.

Even short exposure to temperatures above 50 °C can lead to interference patterns on the surface of aqueous solutions in polystyrene containers. Polypropylene does not melt when autoclaved, although it does become more brittle and may craze after repeated autoclaving. The chemical resistances of these plastics are described in the *Merck Index* as well as in the literature of several manufacturers and suppliers. The Dupont-Sorvall booklet, *Tubes, Bottles and Adapters,* has a particularly extensive and useful description.

The pliable polyethylene is useful in squeeze bottles that dispense water, detergents, ethanol, and other mild solvents. Solutions and biological materials such as nucleic acids, proteins, and viruses can be stored in either hard glass or polypropylene. However, in some instances virus titers fall when virus is stored in polypropylene. In addition, very small quantities of almost any charged substance will bind to glass and plastic. Although siliconizing the container will help, storing submicrogram quantities of nucleic acids, proteins, and the like is risky. Polystyrene is best used only when large-scale work makes cost an important factor and when this plastic has proved to be satisfactory. Both of these conditions apply to polystyrene petri dishes and microtiter plates when they are used for growing bacteria and phage.

SILICONIZING GLASSWARE

Glass surfaces of tubes, glass wool, and so on can be made hydrophobic by coating with silicone. Make a 5% (vol/vol) solution of dichlorodimethyl silane (Sigma) in chloroform. This solution can be stored at room temperature for at least a year without losing its effectiveness. Soak the glass in this solution for 5 min at room temperature and then rinse the glass in water several times and bake at 210 °C overnight. It is usually best to turn tubes and other silicone-treated containers upside down during the baking so that droplets of the solution will not leave large deposits in the bottom of the container. The baking will remove all volatile substances, sterilize the glass, and harden the silicone surface. The silane solution in chloroform may be reused indefinitely.

WASHING PIPETTES

Remarkably, no uniform or best solution seems to exist to the problem of washing and sterilizing pipettes. We have found it easiest to wash all the pipettes sufficiently clean that even the most fastidious enzyme reactions are not poisoned by the pipettes. We also sterilize all of them so that any pipette may be used for any experiment, from bacteriological to enzymatic.

PROCEDURE

1. As pipettes are used, discard them into soapy water.
2. Prerinse the pipettes. Fill a Nalgene pipette washer basket one-half to two-thirds full with pipettes. Drain out the soapy water. Rinse the basket and pipettes with filtered tap water for 5 min in an automatic rinsing container. Our own Waltham tap water must be passed through two commercial filters (Commercial Filters Corporation, Model F15-10) before using in pipette washing. Drain the basket of pipettes *very thoroughly,* shaking them rather vigorously. If this is not done, the water will destroy the cleanser very quickly, turning dichromate from red to green (unusable).
3. Soak the pipettes in a cleansing solution. The classic cleansing solution, dichromate in concentrated sulfuric acid, is still the most satisfactory cleanser. A conveniently packaged dichromate, Chromerge, is commercially available. The metal-free dichromate substitutes have a much shorter lifetime. Very carefully, lower the basket of rinsed pipettes into the cleansing solution and put plastic wrap (Saran Wrap) on top as bubbles start to burst. Soak a minimum of 3 h.
4. Rinse the pipettes. Wearing eye protection, lift the basket above the acid and let the acid dribble back into the container. Shake gently as shown in Figure 7.1 and let it drain until the flow is only a few droplets per min. This should take about 3 min. Quickly put the basket into an automatic rinsing container. Turn on filtered tap water so that the flow rate is sufficiently fast that water will drain every minute or so, but slow enough that all pipettes will completely fill with water. The rinse will be more effective if the basket is shaken a little from time to time during the rinsing.

   Usually, rinsing takes 3–4 h but can be less if you keep checking to make sure the pipettes are filling and draining properly. After rinsing, put the basket of wet pipettes on a sink drain ledge for at least 15 min. Then, to check that the rinsing has been adequate, wipe the entire bottom of the basket with pH paper. Inadequate rinsing or a plugged pipette will yield an acid indication. An option at this point in the procedure is to wash the pipettes with distilled H_2O. Two fillings of the automatic rinser are sufficient to remove material that escaped the tap water filter.
5. Dry and sterilize the pipettes. Load empty pipette cans (VWR, stainless steel) not more than two-thirds full with pipettes. Close the cans and bake at 150 °C overnight to dry and sterilize.

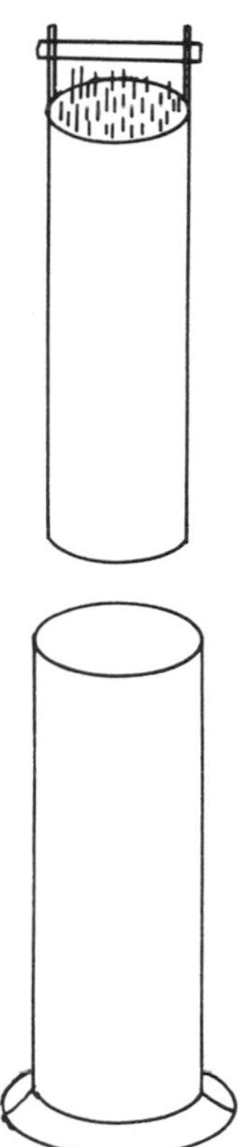
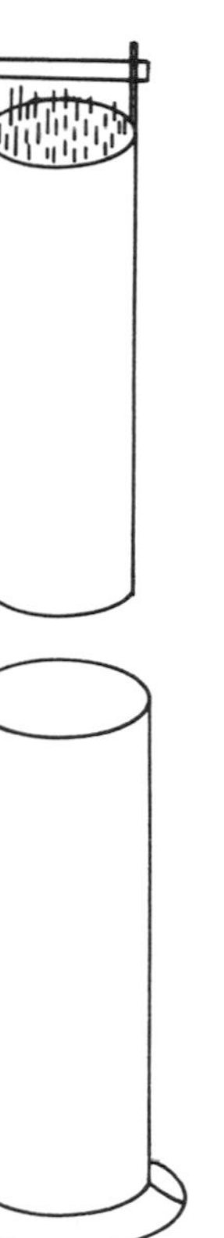

Figure 7.1. Shaking the pipette basket to facilitate complete drainage of the dichromate–sulfuric acid cleansing solution. Shake by rotating.

pH METERS

Most everyday laboratory instruments such as balances are straightforward to operate and come with excellent instruction manuals; hence nothing need be mentioned about them here. However, the instruction manuals accompanying pH meters either omit the following information or bury it so well that many users are unaware of its existence.

Typical pH meters use a combination glass electrode. This consists of an inner pH electrode in contact with the sample through pH-sensitive glass and an outer KCl-containing reference electrode in contact with the sample through a porous glass plug. The pH-sensitive glass and the plug can both become clogged. Clogging causes sluggish response of the electrode as well as decreased sensitivity. Such clogging most often results from protein denatured on the glass. This can be cured by immersing the electrode overnight in a solution of 1 mg/ml pepsin (crude) in 0.1 N HCl. Soaking the electrode for the same amount of time in 0.1 N HCl will overcome less severe clogging. A brief acetone rinse often removes hydrophobic substances.

There are two basic controls available on a pH meter. One regulates the gain of its amplifier, and the other is an offset adjustment. These may be disguised under a variety of names like "temperature," which is a gain control, and "buffer adjust," which is an offset control. There may even be several knobs regulating the same function. Because pH electrodes lose sensitivity with age, the gain control will often have to be increased to allow the pH meter to be accurate at all pH levels. To adjust the pH meter so that it accurately measures all pH levels, note the following facts: (1) the pH meter measures and indicates a voltage produced by the electrode; (2) the gain control changes the amplification of the voltage; and (3) the offset adds a fixed constant to all voltages and pH levels. If the temperature setting required is much different from the ambient temperature, this means the electrode may be clogged. Try cleaning it. If that fails, live with the adjustment. When it is no longer possible to adjust the electrode because it is too insensitive, replace it.

COMMENTS

1. With some weakly buffered solutions, the pH reading is altered by stirring the solution. For an accurate reading measure the pH of an unstirred solution.
2. Some electrodes give false readings with Tris buffers. Sigma Chemical Co. lists them in its Tris bulletin.
3. Some electrodes have a rubber plug at the top. This should be open during use in order to allow pressure equilibration and shut when not in use in order to reduce evaporation of KCl from the reference electrode.

4. Most pH electrodes need to be filled with saturated KCl. Check the level periodically and add more if necessary.
5. Store the probe immersed in saturated KCl. Rinse it with H_2O, but do not wipe off the remaining drop because a static charge may thus be given to the probe. Remove the drop by touching it to the side of the rinse vessel.
6. If excessive KCl crystals have formed in the bottom of the probe (Figure 7.2), dissolve them by warming the probe under running water and replacing the saturated KCl solution with a new saturated KCl solution which has been chilled on ice water. Never add any solution to the probe except the commercial saturated solutions prepared for pH probes. Several heat-chill cycles should suffice to solubilize KCl crystals in a probe.

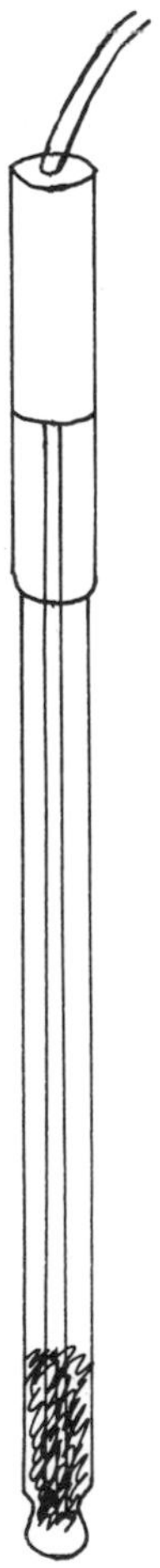

Figure 7.2. A glass combination pH probe with crystals of KC1 on the bottom.

BUFFERS

Tris (tris[hydroxymethyl]aminomethane)

Tris buffers are convenient to make and use, but possess several drawbacks. They chelate some ions. They also possess a rather large temperature coefficient, so it is necessary to be alert to the pH change upon cooling or heating a Tris buffer from room temperature. Most often, and in this book as well, a given pH refers to a measurement made at room temperature even though the buffer may be used at a different temperature. Another drawback of Tris-buffered solutions is that their pH varies considerably with concentration.

Usually 1 M stocks are made by dissolving Tris base in water, stirring in the desired acid, often HCl, until the desired pH is obtained, and then adjusting the final volume to give a concentration of 1 M.

Phosphate

Phosphate buffers are frequently used to buffer enzymatic reactions that require a pH between 6 and 8. As with Tris buffers, the pH varies appreciably with concentration. Thus, phosphate buffers are conveniently made by diluting stocks of monobasic and dibasic phosphate to the desired concentration, then stirring one into the other until the desired pH is attained.

Phosphate-buffered solutions support the growth of microorganisms to densities high enough to generate serious problems from contaminating nucleases, but so low that the contaminating growth cannot be seen by the eye. Hence, caution is necessary. Sodium phosphate stocks of 1 M do not become contaminated with bacterial growth. Potassium phosphate stocks of 0.2 M are near the limit of useful solubility and are best stored frozen. Rapid thawing before use is easily done in a microwave oven.

Good Buffers

A series of buffers originally selected for their nontoxic effects on enzymatic reactions has been described (Good et al., 1966). In our experience these have never poisoned a reaction. They may be obtained from the standard biochemical suppliers.

Cacodylate

This is used to buffer around pH 7. It is useful for long-term reactions because it contains arsenic and inhibits bacterial growth.

BECKMAN ULTRACENTRIFUGES

The manuals for the Beckman ultracentrifuges and rotors are very detailed and have all the information necessary for operation. However, some important information is hard to find among the great mass of information. In addition, there are a few tricks to loading tubes that are not mentioned.

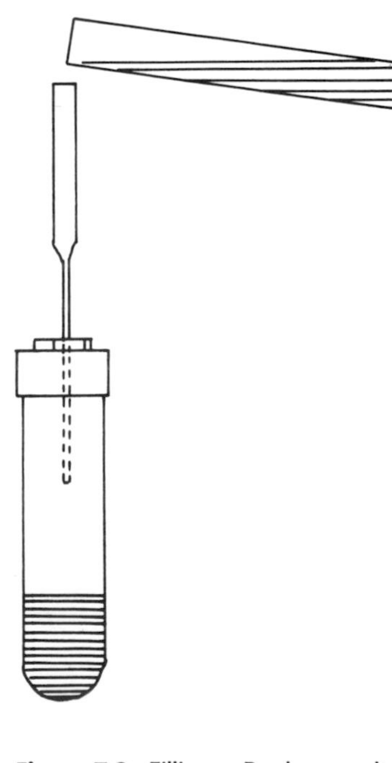

Figure 7.3. Filling a Beckman ultracentrifuge tube. A pipette is often used to pour liquid through the Pasteur pipette funnel.

Filling and Balancing Tubes

PROCEDURE

1. Balance the tubes within 0.5 g. For aqueous solutions the eye is a good estimator, but with denser solutions it is necessary to use a balance.
2. Fill the noncapped swinging bucket tubes to within 4 mm of the top. Thin-wall capped tubes for angle rotors must be completely filled.
3. Tubes with aluminum caps present several problems, some of which will be overcome with the new heat-sealable tubes. For tubes with the aluminum caps, it is best to attach the cap to the tube using the Beckman vise. Check that the tube is well attached by giving it a vigorous yank by hand while the cap is still held in the vise. If it holds then fill the tube using a broken-tipped Pasteur pipette as a funnel in the Allen screw hole of the cap (Figure 7.3). If this funnel is slightly raised, no air traps will occur and 10–20 ml of aqueous solution can be pipetted into the tube in less than 1 min. If the aluminum cap has scratches that are not superficial then throw it out, or it may be found

broken and forced into the bottom of the rotor after the centrifuge run. Balance tubes must be filled with solutions of the same *density* as the sample tubes and then balanced by weighing. Use a 25-ml syringe with an 18-gauge needle to complete filling a tube with mineral oil. Screw the Allen screw into the top and invert the tube. There should not be much more than 0.1 ml of air bubbles. More air in the tube will allow the tube to collapse and perhaps even pull away from the cap, ending as a shriveled pile at the bottom of the rotor hole.

Loading Rotors

PROCEDURE

1. A 12-place rotor can be balanced with 2, 3, 4, 5, 6, 7, 8, 9, 10, or 12 equally filled tubes (Figure 7.4). For swinging bucket rotors Beckman now recommends the use of all buckets whether or not they have tubes inside. Always put the buckets on the hanger that has the same number and never use buckets from a different rotor. The problem is that after a rotor and bucket have been used together they deform to make a uniform surface contact that is unique to each bucket and its slot shoulder. A differently deformed bucket may only make a few contacts with a strange slot shoulder, and this will cause a severe strain on the metal.

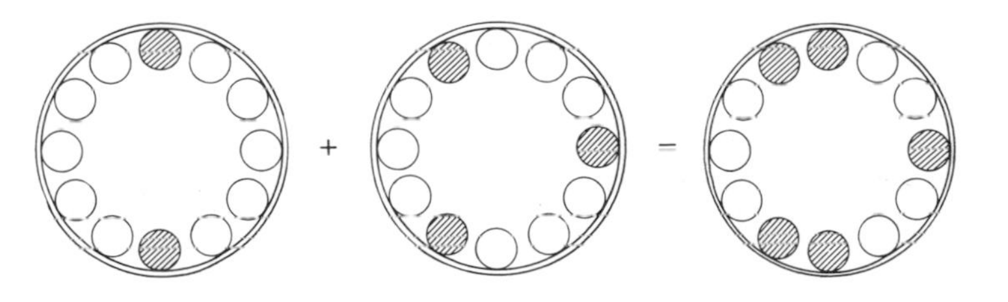

Figure 7.4. How 5 tubes in a 12-place rotor may be considered as a combination of 2 oppositely placed tubes plus 3 tubes at corners of an equilateral triangle.

2. Samples must be isolated from the vacuum of the centrifuge chamber or they will evaporate and unbalance the rotor. Check all O-rings and gaskets, including the O-ring in the angle rotor handle. They should be slightly greasy and not gritty. When tightening screws to form seals, do not tighten to the point that metal-to-metal contact is made.

DRAWING FIGURES

Satisfactory figures may be drawn with little practice, and usually with less bother than explaining exactly what is wanted to a draftsman.

Minimum equipment is the following: an 18 × 24-in. wood drafting board; a 24-in. T square; an 8-in. triangle; a French curve; a template for making various-sized circles, squares, and triangles; and a Leroy lettering set with #0, 1, 2, 3, and 4 pens and 120CL, 200CL, and 290CL templates. India ink, tracing paper (Albanene paper from Keuffel and Esser is good), and masking tape are also needed.

Begin by making a full-sized drawing which can then be traced. It is best to make these drawings at least twice the size they will be when in their final published form. Reduction will sharpen the drawing and eliminate roughness. Be aware, however, that the thickness of lines, the size of letters, numbers, and symbols, as well as all the spacings will also be reduced. Note, also, that a publication will be much more attractive if the style and proportions of all drawings are uniform. For this reason, it is usually best to make all drawings at the same magnification relative to their publication size and to use the same general style and the same style and size of lettering.

After making the drawing, place the tracing paper over the drawing and tape both to the board. Usually it is easiest to ink in the borders, then the data points, then the actual curves, and finally the lettering. Note that the bottom edge of the T square is slightly recessed so that it does not touch the paper and hence ink tends not to flow under it. The same is not true of most triangles and curves. If the edges are not recessed, the pen must be slightly inclined so that its point does not touch the edge. Typically, the templates for drawing data points are too thin for such tilting maneuvers to be successful. For these it is best to apply small pieces of masking tape to the bottom of the template so as to lift it off the paper, like the edge of the T square.

Lettering is the hardest part. Begin by lettering a line or axis identification on a scrap of paper. The spacing scale on the templates is worthless. In order to properly space letters, arrange that the open *area* between the actual ink of the letters is the same. Thus, the distance of closest approach between O and Q is smaller than the distance of closest approach between an N and an M. Once a satisfactory line of lettering has been made, it can be centered in its proper location under the drawing and traced.

Lettering for figures often can be done on a typewriter. An IBM Selectric permits insertion of a typing ball with sans serif type that is similar to the lettering produced by a Leroy template. The required size of the lettering in the final reproduced figure then dictates the size of the figure. To rescale a figure so that lettering may be typed, it is convenient to use a pair of proportional dividers. These may be obtained from most art or drafting supply houses. Corrections of minor mistakes can be made with a razor blade. It is also possible to use a motorized eraser and a thin steel erasing guide.

SLIDES AND NEGATIVES

As in the case of drawing figures, it is often easier to make prints or slides exactly as you want them rather than to have a professional photographer do the work. Photographic processes are also an essential part of many experiments, so the necessary materials are usually readily available.

Polaroid Slides

Polaroid makes two types of slide film that we find useful: 146L is a high contrast film for line drawings; 46L has normal contrast and should be used for subjects with intermediate gray tones.

PROCEDURE

1. Put the drawing under the Polaroid copy camera on an illuminated stand and cover it with a piece of clean glass to keep it flat. Turn on the copy stand lights. Turn off the room lights to eliminate spurious reflections.
2. Center the drawing by looking at the image on the ground glass. Adjust the image size by changing the height of the camera. Focus with the lens aperture and shutter wide open.
3. Move the film holder into position. Set the f-stop and shutter speed. Then pull out the sliding film mask, expose the film, and develop it. For 146L film, using a copy stand with four 150-W flood lamps, try an exposure of 0.5 sec at f/22; for 46L try an exposure of 0.25 sec at f/16. If the slide is overexposed, that is, if it is too light, then reduce the exposure. If it is too dark, increase the exposure.
4. Polaroid slides are normally fixed with Polaroid Dipit solution which is expensive, deteriorates fast, and streaks the film, but does fix slides in only 20 sec. It is also possible to fix Polaroid slides with normal photographic fixer: fix for 5 min; wash in running water for 10 min; rinse in distilled water; and then dry in air.

Conventional Slides and Negatives

The products of conventional photographic methods require more labor than do Polaroid films, but they are of higher quality. Also, conventional methods allow much more room for optimizing to increase contrast, detail, and so on. The many varieties of film, paper, and solutions available from Kodak, as well as their characteristics when used in various combinations, are concisely summarized in the *Kodak Dark-*

room Dataguide, available at almost all photography stores.

When making negatives to be used for making prints, high-quality results can be obtained with 4 × 5 Kodalith film developed in a high-contrast developer (Kodalith developer; D-11 or D-19) for the time given in the developer instructions. Use only a dark red safelight. The 4 × 5 film holders fit a Polaroid copy camera and all the instructions on using the camera for Polaroid slides apply. Make a series of test exposures on a single sheet of film by pulling the sliding mask out in steps between exposures. Examine the processed test sheet with a bright light and choose the shortest exposure that gives the maximum darkness on the film. Overexposure makes the dark areas expand into the light areas, whereas underexposure leaves many light pinholes in the dark areas. Touch up pinholes and draftsman's mistakes with photographic opaque. Make positives by contact reproduction or by projecting while the negative is mounted in a Polaroid frame or between glass.

Several types of 35-mm film may be used instead of 4 × 5 film. The 35-mm film is more convenient since a darkroom is unnecessary, but touching up the negative is more difficult. Although high contrast copy film reproduces line drawings well, adequate results are obtained with Kodak Tri-X, Plus-X, or Panatomic-X. The latter three are more versatile, because they make good reproductions of subjects with gray tones. Note that the higher ASA of Plus-X and the still higher ASA of Tri-X are achieved at the cost of increased grain size and therefore lower resolution of detail.

One high contrast 35-mm film, SO-185 (Kodak), is good with line drawings and adequate with gray tones. It has the added advantage that developing in D-19 gives a positive rather than a negative transparency. Thus, if slides are the only objective, it is usually the film of choice.

Generally, it is best to use a test roll of 35-mm film to determine the combination of exposure length and f-stop which yields adequate results. If no darkroom is available,

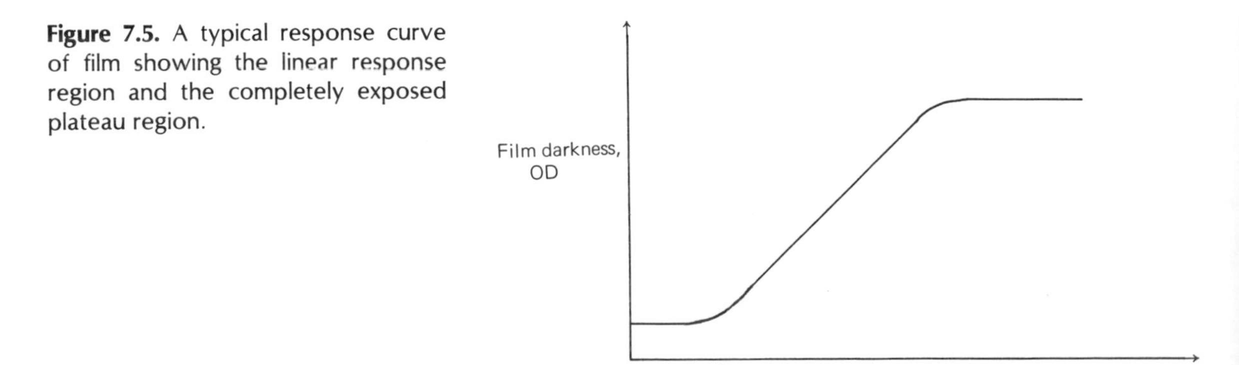

Figure 7.5. A typical response curve of film showing the linear response region and the completely exposed plateau region.

after exposing the film transfer it to a film-developing tank using one of the commercially available dark bags. Then develop and fix according to the directions that come with the film. Transparencies may be mounted in the Emde 2 × 2 aluminum slide binders, with 35-mm foil masks, or somewhat more easily with plastic GEPE glass or glassless diabinders. The negatives of line drawings make acceptable slides, but for more standard positive slides use either the SO-185 film or rephotograph the negatives. A slide copying attachment on the 35-mm camera is most helpful for this final step.

FILM SENSITOMETRY

A representative graph of the OD of developed film plotted against the logarithm of the exposure is shown in Figure 7.5. The slope (known as the gama) of the linear portion of this curve depends on the type of film, the type of developer, the time and temperature of development, as well as the amount of agitation during development. For quantitative experiments you should determine this curve for your system and use the linear portion of the curve for your experiments. Use approximately the same exposure time for making the standard curve as you expect to use in your experiments, because, for very long or very short exposures, the OD of the developed film varies with the exposure time as well as with the total amount of radiation received by the film. This exposure time effect is called reciprocity failure.

To obtain consistent development of photographic emulsions, you must use fresh developer. After it has been mixed with water, the developer will stay fresh for several weeks and often longer, provided that it is kept in the dark in a tightly closed bottle with only a small amount of air. See the instructions that come with the developer for details of the storage properties of particular products. Agitation during development is also important to consistent developing results. This means that you should use a standardized routine to rock the developing tray or to stir the developer several times a minute during development. Also be sure that the sheets of film or paper do not stick together during development. Stop the development by rinsing for a few seconds in 2% acetic acid and then fix the film for a standardized amount of time, usually about 5 min.

Many of these recommendations are superfluous unless you are doing an experiment in which you want to get information from measurements of the OD of the film, as, for example, in quantitating autoradiograms. For less critical experiments, use old developer at any convenient temperature and rinse with water instead of acetic acid.

AUTORADIOGRAPHY AND FLUOROGRAPHY

Photographic emulsions are sensitive not only to photons, but also to ionizing radiation. Thus, they can be used to locate the positions of radioactive spots following chromatography or electrophoresis. ^{3}H, ^{14}C, ^{35}S, and ^{32}P may all be detected. Frequently it is useful to increase the autoradiographic sensitivity to ^{32}P decay. This can be done by using a fluorescent screen in a sandwich, as shown in Figure 7.6. The high energy electron given off in the decay of the phosphorus nucleus passes through the photographic emulsion, leaving the usual trail of crystals that will later be reduced. If a fluorescent intensifying screen is placed on the other side of the film, then when one of its crystals is struck by a high-energy electron it will give off many photons which, in turn, will sensitize many crystals in the emulsion. Useful parameters in this method have been explored by Swanstrom and Shenk (1978). The film sensitivity can be increased 10-fold by this fluorographic method. Note that different emulsions possess different color sensitivities and different fluorescent screens emit photons of different colors. Thus, some combinations are much better than others. For an exposure longer than a few hours, it is usually better to freeze the entire sample-film-screen sandwich at −70 °C. This prevents diffusion of the sample, but more importantly, it stabilizes sensitized crystals in the emulsion and thereby increases the sensitivity of detection.

Below, we describe a protocol for the autoradiography of a filter paper containing ^{32}P. This protocol is used for Southern transfers (p. 152), colony and plaque hybridization experiments (p. 147), and many other standard methods. Fluorography of ^{3}H, ^{14}C, and ^{35}S embedded in gels is accom-

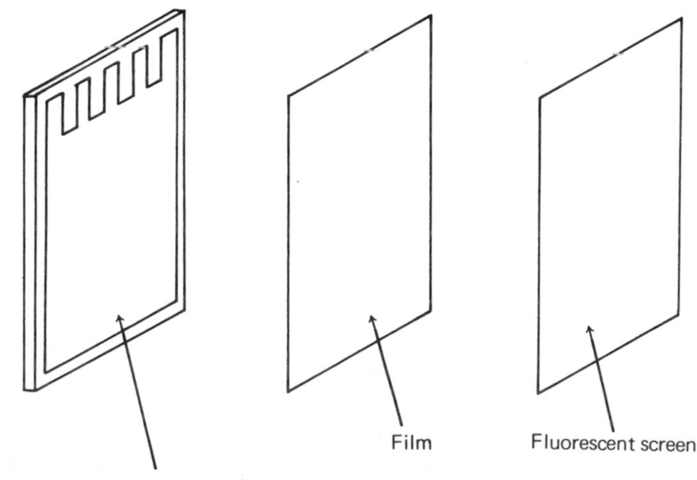

Figure 7.6. Exploded version of a sandwich for enhancement of an autoradiograph showing placement of the fluorescent screen on top of the film.

plished by impregnating the gel with a fluorescing scintillant. This procedure is described on p. 87 (Acrylamide Gels for Proteins).

PROCEDURE

Blot the filter paper to damp dryness on paper towels and then place it between two layers of Saran Wrap. This Saran Wrap sandwich keeps the filter damp so that the filter can be handled without cracking. The sandwich also allows photographic film to be placed next to the damp filter without the film becoming wet and therefore useless. Use the sandwich to expose Kodak XAR 5 x-ray film utilizing Dupont Cronex Lighting-plus XG intensifying screens at −70 °C. Plastic lead-backed screen cassettes (AQ; x-ray exposure holders) are convenient for this. Place the film on the intensifying screen and the Saran Wrapped filter paper on the film. Use a Kodak GBX safelight with a 15-W bulb when setting this up. Put 2-mm thick sheets of aluminum on both sides of the cassette and tape the sheets together. This will help to shield radiation from neighboring autoradiograms in the freezer and also assures good contact between film and paper.

COMMENTS

1. Air bubbles between the filter and the Saran Wrap can expose the film with characteristic "crow's foot" patterns. Minimize the air bubbles by taping Saran Wrap to a flat surface, putting the filter on it, and then laying a Saran Wrap sheet on top while holding it under tension with your hands. The "crow's feet" seem to be a problem of static electricity. Some people find this problem can be eliminated by wiping the surface of the Saran Wrap with a damp paper towel and letting it dry before putting that surface against the x-ray film.
2. Intensifying screens are also phosphorescent, so do not expose them to light before constructing an autoradiographic sandwich.

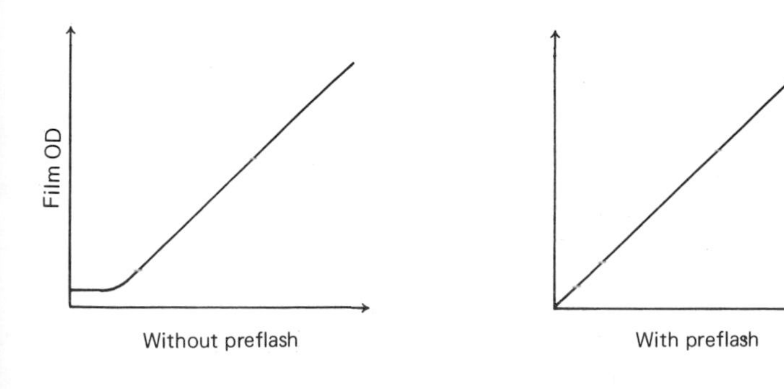

Figure 7.7. The effect of preflashing film upon its sensitivity.

3. To increase sensitivity for the faintest bands, it is beneficial to "preflash" the film. Ideally, this preflash places the film just at the bottom of the linear portion of the response curve, as shown in Figure 7.7. The preflashing is conveniently done by using a printing enlarger to expose the film before autoradiography. The best preexposure can be determined empirically, and seems to be that which produces an OD of about 0.1 in the developed film. This OD may be determined by placing the developed film in the light path of a standard spectrophotometer and measuring at a visible wavelength.

DIALYSIS TUBING

Dialysis is one of the traditional methods for changing the buffer or salt of a protein or DNA solution. Despite the virtues of Sephadex columns for achieving similar ends, dialysis is often still the method of choice. Dialysis tubing is not just "dialysis tubing." It is available in a wide variety of diameters, wall thicknesses, and pore sizes. For extensive information on the tubing available, contact the Food Products Division of the Union Carbide Corporation. Note also that very large savings are possible if the tubing is purchased in 1000-ft rolls of random lengths rather than in 100-ft continuous rolls. Dialysis tubing is identified by its diameter in inches measured in 32nds of an inch. Thus, tubing with an inflated diameter of 0.25 in. is usually identified by the manufacturer as #8 dialysis tubing. Below is a summary of the properties of dialysis tubing most commonly used in molecular biology:

Size	Wall thickness	Apparent pore size
8	thick	medium
20	thin	large (do not use with proteins of molecular weight below 5000)
23	medium	small
27	medium	medium

Dialysis tubing usually is filthy when it arrives. It is safest to wash the tubing extensively before any valuable biochemical is entrusted to it. Also remember that the cellulose from which dialysis tubing is made is tasty to some microorganisms. For this reason it is prudent to prepare new tubing at about 6-month intervals and to occasionally check the sterility of the water in which dialysis tubing is stored. This check is easily done by streaking the water on a YT plate.

WASHING PROCEDURE

1. Cut the tubing into 20 pieces about 40 cm in length.
2. Boil the pieces for 15 min in 1400 ml water plus 10 ml of 0.2 M EDTA.
3. Repeat step 2 with a fresh EDTA–water solution.
4. Boil for 15 min in 800 ml double-distilled water plus 5 ml of 0.2 M EDTA.
5. Boil for 20 min in 800 ml double-distilled water plus 0.8 ml of 0.2 M EDTA.
6. Store in a refrigerator in 800 ml double-distilled water plus 0.8 ml of 0.2 M EDTA.

Boiling is most effective if the tubing is held under the surface with a large flask partially filled with water (Figure 7.8).

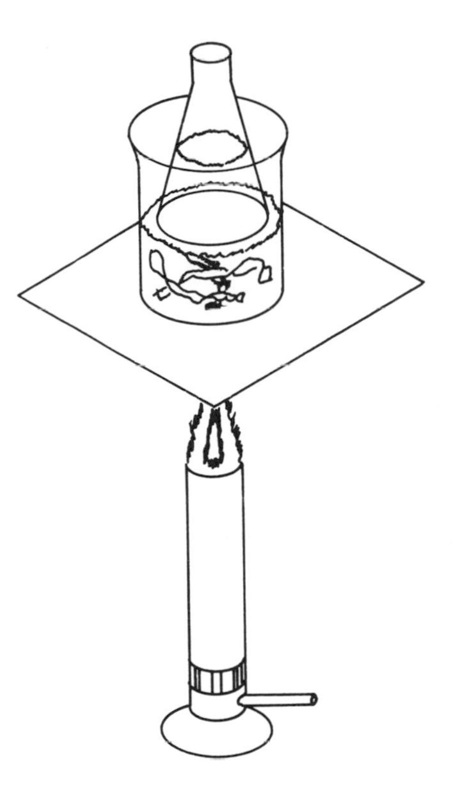

Figure 7.8. Keeping dialysis tubing in boiling water with a flask during boiling.

Pinhole leaks are not unknown and it pays to test a dialysis sack at the last moment before tying the final knot. A light squeeze of the sack while its untied end is twisted shut will usually reveal any leaks.

DISTILLING PHENOL

Phenol extraction is an excellent method for deproteinizing DNA because it is rapid and nearly foolproof. However, the oxidation products of phenol are reputed to produce cross-links between DNA strands. Thus, many laboratories repurify phenol by distillation. This is most conveniently done in sizeable quantities and then aliquots of the phenol are stored frozen.

MATERIALS

1. 2000-ml round-bottom flask with ground-glass taper joint
2. Heating mantle for flask
3. Distilling adapter with top joint for thermometer
4. Thermometer with ground-glass joint
5. Condenser, about 18 in. long
6. Variable voltage source, for example Variac
7. Glass wool to pack around top of flask

PROCEDURE

Set up the distilling apparatus as shown in Figure 7.9 in a chemical hood. Add 200 ml of water to 2 kg of phenol and melt in a 65 °C water bath. This melting can be avoided if you purchase Liquefied Phenol, which is 10% water and is liquid at room temperature. Pour the phenol into the distilling flask using a funnel. Add approximately 50 boiling chips.

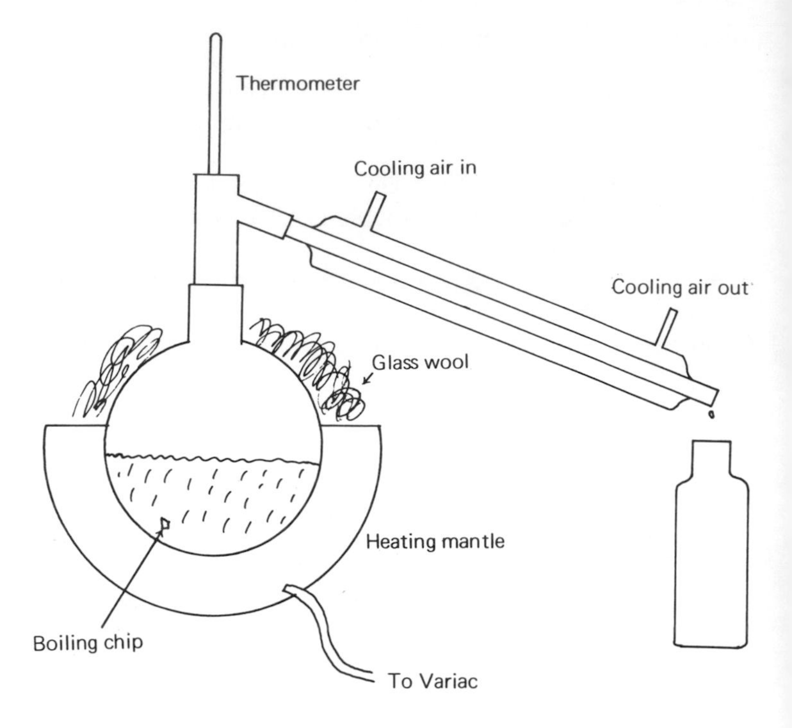

Figure 7.9. Setup for distilling phenol.

Adjust the air flow to achieve a moderate flow through the condenser. Begin heating. Pack the glass wool around the top of the flask. Begin collecting the distilled phenol when the thermometer indicates about 160 °C. It is most convenient to collect into a 3-liter bottle containing 200 ml of water. The water greatly lowers the melting temperature so that the phenol may later be poured into milk dilution bottles and then stored frozen at −20 °C. The water in the phenol will allow it to be melted in a 37 °C water bath. Phenol is stable for years when stored at −20 °C. Some fiddling with the voltage to the heating mantle and the air supply is necessary to prevent phenol crystals from plugging the outlet of the condenser and to ensure that all the phenol vapors are condensed by the condenser. Be aware that the residue from the distillation is liable to be highly explosive and the distillation should be stopped when about 100 ml remain in the flask. At this point the apparatus should be disassembled and the residue flushed down the sink.

RECOVERING USED CsCl

CsCl from ethidium bromide density gradients and from other types of gradients can be recycled for use in other preparative centrifugations. The product of this recycling procedure has been used successfully for the plasmid and DNA preparations described in this book. If particularly finicky uses are planned for the CsCl, it is probably wise to substitute one or two crystallization steps for the fourth step in this procedure.

PROCEDURE

1. Put the used CsCl in a Pyrex baking dish and bake it overnight at 600–700 °C. A typical glass blower's furnace is convenient for this, but these furnaces are usually not ventilated. Since used CsCl usually contains a variety of organics it is wise to do this baking in a well-ventilated room at a time when no one will occupy it. The Pyrex dish will break during the baking but almost all of the solid CsCl can be recovered.
2. Dissolve the CsCl in water and filter it through charcoal (Norite). This should remove all colored material in the solution. If it does not, refilter the solution through more charcoal.
3. Filter the solution through Whatman #1 filter paper.
4. Boil off most of the water and then allow the CsCl to dry.
5. Grind the CsCl in a mortar and pestle until it has a fine granular texture.

SOURCES OF CHEMICALS

Fisher and J. T. Baker chemical companies provide reliable standard chemicals. Some sources of unusual chemicals are listed below together with comments that summarize our experience with these sources. In the long run, it usually saves time to purchase the highest purity chemical offered in the catalogs. To save money, in some procedures that have been tested, such as TCA precipitation of nucleic acids, we use the impure technical or practical grades. In the protocols of this book we specify the supplier and/or the grade of the chemical when we have found that they matter or when we suspect that they matter.

ORGANICS

In general, one must be skeptical of the purity of organics, since they frequently are not carefully assayed.

1. Eastman Organic Chemicals (Rochester, N.Y.) offers several grades of chemicals that seem to be of reliable quality. White labels indicate good quality. The yellow and blue labels are used to indicate fair- to poor-quality chemicals.
2. Aldrich Chemical Co. (Milwaukee, Wis.) offers some ultrapure chemicals at reasonable prices.

3. Schwarz/Mann (Orangeburg, N.Y.) is a reliable source of ultrapure sucrose, ammonium sulfate, and urea.
4. Columbia Chemicals (Columbia, S.C.) offers a good selection that includes many fluorine-containing chemicals.
5. P. L. Biochemicals (Milwaukee, Wis.) over a long period of time has been the most reliable supplier of nucleoside triphosphates and other nucleic acid precursors.

INORGANICS

Gallard Schlessinger (Carle Place, N.Y.) is currently an inexpensive source of CsCl that is suitable for preparative and most analytical equilibrium density gradient centrifugations.

HAZARDS AND CAUTIONS

The following collection of laboratory safety principles is certainly not comprehensive. It does not, for example, list all dangerous chemicals. A number of large and authoritative safety monographs already exist and we have little if anything to add to them. Our objective is to describe some general and specific principles of safety that we follow when working in our laboratories. These principles deal with common chemicals and operations used in a molecular biology laboratory. By and large we have adopted most of the principles because laboratory folklore suggests that we should. Luckily we do not have direct experience that leads us to know that all of these principles are necessary. Note that we do not repeat the cautions that have already been given in the protocols of this book.

Desiccators, Preparative Centrifuges, and Vacuum Pumps

The conventional desiccator is *not safe* under 0.5 atmosphere. At all times it should be used with a commercially available steel-guard cage. The implosion of a vacuum desiccator is operationally indistinct from an explosion. Taping a desiccator for safety is just wishful thinking. A good substitute desiccator for desiccating small volumes is a test tube fitted with a 3-way stopcock.

Unarmored centrifuges such as the Sorvall SS-1 are *not safe*. It is probably asking too much not to use them, but they should always be operated away from people.

Mechanical vacuum pumps often become very hot when used for several hours and have been the cause of many laboratory fires. It may be wise to use a smoke detector if you are going to run a pump overnight.

Chemicals

If you need to use an unfamiliar compound, it is always wisest to read about it in the *Merck Index* before using it. The *Merck Index* is inexpensive and no laboratory should be without one. The National Academy of Science (U.S.A.) has recently published an excellent report, *Prudent Practices for Handling Hazardous Chemicals in Laboratories.* It can be obtained from the National Academy Press, Washington, D.C. Another very useful, although more expensive source of information about the hazards of almost all chemicals is Sax's *Dangerous Properties of Industrial Chemicals* (1979). In general it is more detailed than the *Merck Index.* It also includes information about reasonable methods for safely storing and shipping chemicals. In the United States other useful sources of information, particularly in emergency situations, are the poison control centers, whose telephone numbers are usually found on the inside cover of telephone directories.

Almost all organic solvents are highly flammable. The more volatile ones, such as ether, benzene, low-boiling petroleum ether (ligroin), and carbon disulfide, are easily ignited by flames, cigarettes, or sparks. An ordinary noninduction electrical motor and the thermostat of a water bath are able to ignite them. The ignition temperature of CS_2 is noteworthy, being less than 100 °C; it will ignite on contact with a steam line. The ignition temperature and explosive limits of other compounds can be found in any standard chemical safety book. When working with flammable solvents be sure that ignition sources are well removed. The vapors of these solvents are usually heavier than air and will flow down a bench top to find a flame 5–10 ft away.

Most organic compounds are relatively harmless if not ingested. There are, however, numerous exceptions. Some of the more common exceptions are listed below.

1. Carbon tetrachloride, chloroform, and other "halocarbons." These are usually nonflammable, but when in a fire they do give off the poisonous gas, phosgene. These compounds cause damage to the liver and should be used in a chemical hood. When working with small quantities, a well-ventilated area is satisfactory. *Never* evaporate them into the laboratory air.
2. Methanol. This is modestly flammable, but toxic and should not be inhaled wantonly. It is absorbed reasonably well through the skin. Use it in a hood or, with small quantities, in a well-ventilated space.
3. Acetone. This is flammable but not particularly poisonous. However, it should be treated with respect.
4. Benzene, toluene, and xylene. These are highly flammable and their vapors are injurious to the liver. Fire is the most frequently encountered disaster with these,

although continual breathing of the vapors will certainly create internal problems.

5. Acids. Organic acids are flammable. The toxicity of acids is usually not a problem. Hydrofluoric acid is dangerous. If you are not well versed in its properties do not use it. When using sulfuric acid make sure to add it to water and not vice versa.
6. Phenylhydrazine. This chemical is frequently used to characterize sugars. It is quite toxic and readily absorbed through the skin.
7. Dimethylsulfate. This is easily absorbed through the skin and very toxic. Ammonium hydroxide solution should be kept handy for application to any dimethylsulfate that is spilled on the skin.
8. Dimethyl sulfoxide, dimethyl formamide, and acrylamide. All these are flammable but relatively innocuous, except for acrylamide which is reported to be a neurotoxin. Care should be exercised to avoid skin contact and inhalation of these chemicals. Be aware that dimethyl sulfoxide greatly facilitates the passage of chemicals through the skin. Thus, the combination of two relatively harmless chemicals, dimethyl sulfoxide and a water-insoluble organic, can become highly toxic.
9. Nitrosoguanidine and other mutagenic reagents are usually potent skin irritants and carcinogens.
10. Dansyl chloride, phenyl isothiocyanate, and related reagents for modifying proteins. These should be kept from contact with the skin. Chemicals of this general type are, by definition, reactive toward biologic molecules (i.e., laboratory workers). If nonvolatile, it normally suffices to keep them off the skin. If volatile, ingestion of the vapors must be considered and guarded against.

Storage of most chemicals presents no problem. Strong oxidizing agents such as hydrogen peroxide, ammonium persulfate, and concentrated nitric acid should not be stored in the immediate vicinity of reducing reagents such as most organic compounds.

Compounds such as hydrogen peroxide that decompose to a gas should *never* be stored in a refrigerator with a mechanical latch. Flammable solvents should not be stored in bulk in a laboratory. They constitute a definite fire hazard.

Electrical Hazards and Protections

The passage of more than 10 mA through your chest can be lethal. The voltage required to drive this amount of current through a human depends on the resistance of the connections and can be as high as several thousand volts since the conductivity of the skin is low. However, if a good electrical connection is made, for example through sweaty palms or a

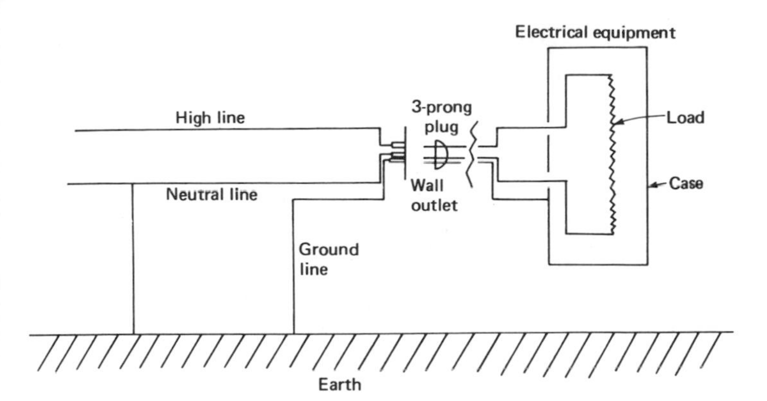

Figure 7.10. Grounding of electrical equipment. The figure shows electrical power provided on two lines, the high line and the neutral line, which is grounded. The case of the electrical equipment is grounded via the ground line.

hand immersed in a salt solution, tens of volts will electrocute you.

Two wires suffice to provide electrical power to equipment (Figure 7.10). For efficiency in electrical distribution, one of these wires or lines is connected to ground and is called neutral while the other is hot or high. Thus, the electrical potential existing between the neutral line and ground, that is, water pipes or other conductive materials in electrical contact with the earth, is at all times near zero. A hazard of a shock exists if a connection is made between the two power lines or between the high line and anything that is grounded.

A third line, the ground line, is used to reduce the dangers of electric shock resulting from shorts within equipment. Such a short could connect the case of the equipment to the high side of the power line. Then a hazard of shock would exist if you touched the equipment and at the same time made contact with anything that was grounded. The danger is reduced by making a separate connection between the ground line and the equipment case. Thus even if a short does occur, the case of the equipment cannot become "hot" and shock you if you also make electrical contact with the ground.

Additional electrical protection is provided by ground fault interrupters. These devices measure the current provided to the load, e.g., a motor, from the high line and compare it to the current returned from the load on the neutral line. Should there be any difference in these two currents, the ground fault interrupter very rapidly switches the current off. Thus, if a path of current flow has developed which bypasses the neutral line, i.e., goes from the hot line through you and on to a water pipe, the ground fault interrupter rapidly turns the current off. The turnoff is rapid enough that the possibility of being severely shocked is much reduced. Ground fault interrupters are particularly valuable in situations in which you may be grounded and likely to touch electrical equipment which if faulty may be "hot." Cold rooms, laboratories, bathrooms, and outdoor construction sites are areas where ground fault interrupters are particularly valuable.

In addition to the obvious precautions one should take with electrical equipment, one should also be aware of the dangers posed by electrical capacitors in some equipment. Large electrolytic capacitors which exist in some power supplies used in electrophoresis and in most power supplies for digital equipment, that is, computers, can be quite dangerous. Often the charge remains even though the equipment is switched off. Surviving victims of this type of shock generally adopt the practice of shorting such capacitors with a wire before working on the equipment.

Appendix I

Commonly Used Recipes

RECIPE LIST

1. Stock solutions for media preparation
2. YT and LB media and plates
3. TB plates for lambda phage
4. A plus B minimal medium for plates
5. M10 liquid minimal medium
6. Phage R17 and P1 plates
7. Phage P1 top agar
8. Phage T4 plates
9. Phage Mu plates
10. Tetrazolium plates
11. EMB and EMBO plates
12. Xgal plates
13. Arabinose fucose minimal plates
14. Ribitol plates
15. Stab-agar for storing strains
16. Phage lambda suspension medium and storage buffer
17. TM medium for nitrosoguanidine mutagenesis
18. 50× TE buffer for DNA
19. 2× freezing medium for cells
20. 20× SSC buffer for DNA
21. Formamide
22. Carrier DNA
23. Distilled water
24. Chocolate chip cookies

1. Stock Solutions for Media Preparation

Sugars are made at 20% (wt/vol).
Amino acids are made at 1%.
Vitamin B_1, thiamine, is made at 1 mg/ml.
$MgSO_4$, $CaCl_2$, and $MnCl_2$ are made at 1 M and $FeCl_3$ at 0.01 M.

EDTA (ethylenediaminetetraacetic acid) must be neutralized to a pH near 7 in order to be dissolved at concentrations up to 0.2 M. To neutralize 100 ml of 0.2 M, approximately 6.5 ml of 10 M NaOH or KOH will be required. It is useful to have both Na-EDTA and K-EDTA available.

These solutions should all be sterilized by autoclaving. If they are in the convenient milk dilution bottles (Fisher 2-943-5), 10 min of autoclaving is sufficient. Generally, several ml of water evaporate during the autoclaving and this amount may be restored afterwards. Mark the meniscus with a magic marker before autoclaving and later add sterile water to restore the former level.

2. YT and LB Media and Plates

These yeast extract–tryptone broth media are two of the most useful media for growing bacteria. Almost all *E. coli* mutants grow on them. These two commonly used media differ slightly in the amount of yeast extract and are probably interchangeable in all cases. Most *E. coli* strains will attain densities near 1 g/liter of medium when grown on YT. Bacterial yield is roughly proportional to the amount of yeast extract and tryptone added up to a level about 5 times YT.

Usually between 20 and 30 ml are poured in 100 × 15 mm plastic petri plates. In humid weather these plates, like all others, require "curing" for a day or two until faint wrinkles appear on the surface. Absolutely smooth plates have a tendency to perspire when placed in the incubator. If, as is frequently the case, the plates must be used immediately, the curing may be hastened by leaving the lids off during the 20 min required for the agar to harden. Airborne contamination is rarely a problem if subsequent incubations need not exceed about 3 days.

To prepare a set of plates, each with different ingredients or different concentrations of an ingredient, it is usually easiest to put the concentrated ingredient in the empty plate, pipette in 25 ml of the medium plus agar containing any other desired components, and mix with the end of the pipette.

	For 1 liter	For 15 liters (1.33-strength)
NaCl	5 g	100 g
Bactotryptone (Difco)	8 g for YT, 10 g for LB	160 g
Yeast extract (Difco)	5 g	100 g
Distilled H_2O	1 liter	15 liters
Dow Antifoam A	–	2 ml

For plates, add 15 g agar (Difco) per liter.

Plates should not be more than 2 days old for plating phage, and, if used for plating $\phi 80$, the NaCl concentration should be lowered 10-fold.

Instead of cluttering shelves with many varieties of top agar, we use TB top agar on YT plates.

3. TB Plates for Lambda Phage

The glucose which is present in the yeast extract of YT or LB medium inhibits synthesis of the lambda phage receptor protein which is located on the surface of the cell. Thus, although YT plates are satisfactory for plating most mutants of lambda, some lambda mutants require the use of TB plates.

Bactotryptone (Difco)	10 g
NaCl	2.5 g
$MgSO_4$	1 ml of 1 M
Distilled H_2O	1 liter
Agar (Difco)	11 g if poured in plates, 8 g if used for top agar

Top agar is conveniently sterilized in milk dilution bottles and kept solid. It is remelted in a microwave oven and cooled to 48 °C before use.

4. A Plus B Minimal Medium for Plates

This is a basic minimal medium to which the desired carbon source, amino acids, vitamins, and antibiotics may be added.

Flask A (1-liter flask)		Flask B (2-liter flask)	
Na_2HPO_4	7 g	Agar (Difco)	15 g
KH_2PO_4	3 g	Distilled H_2O	500 ml
NH_4Cl	1 g		
Na_2SO_4	0.8 g		
Distilled H_2O	500 ml		

a. Autoclave.
b. Mix flasks A and B.
c. Add sugar to 2 g/liter (10 ml of 20%), 10 ml of vitamin B_1 at 1 mg/ml, and desired amino acids at 20 μg/ml.

5. M10 Liquid Minimal Medium

This is a good liquid minimal medium. It differs from the classical M9 medium (Anderson, 1946) in the inclusion of Mn^{2+} which is required by the enzyme, arabinose isomerase.

Na_2HPO_4	7 g
KH_2PO_4	3 g
NH_4Cl	1 g
NaCl	0.5 g
$FeCl_3$	0.1 M, 0.3 ml (shake before using, as this reagent settles after autoclaving)
Distilled H_2O	1 liter

Autoclave, and when cool add the following sterile solutions:

$MgSO_4$	1 ml of 1 M
$CaCl_2$	1 drop of 1 M
$MnCl_2$	add to make 5×10^{-5} M

5a. 20 × Concentrated M10

Na_2HPO_4	140 g
KH_2PO_4	60 g
NH_4Cl	20 g
NaCl	10 g

Add 900 ml H_2O, mix, and filter through Millipore 0.45-μm filter (47 mm diameter). Then add the following:

$FeCl_3$	0.6 ml of 0.1 M
$MnCl_2$	1 ml of 1 M
$CHCl_3$	5 ml, for sterility of the stock

After autoclaving a 20-fold diluted portion, add 1 drop of 1 M $CaCl_2$ (optional) and 1 ml of 1 M $MgSO_4$ (not optional) per liter.

6. Phage R17 and P1 Plates

Agar (Difco)	15 g
NaCl	5 g
Bactotryptone (Difco)	8 g
Yeast extract (Difco)	1 g
Distilled H_2O	1 liter

Autoclave, and when cool add the following sterile solutions:

$CaCl_2$	1 ml of 1 M
Glucose	2.5 ml of 20% (wt/vol)

7. Phage P1 Top Agar

Bactotryptone (Difco)	8 g
NaCl	5 g
Agar (Difco)	6.5 g
Distilled H_2O	1 liter

Just before use, add $CaCl_2$ to make 0.025 M.

8. Phage T4 Plates

	Bottom Agar	Top Agar
Bactoagar (Difco)	12 g	6 g
Bactotryptone (Difco)	13 g	10 g
NaCl	8 g	8 g
Na^+-citrate	2 g	2 g
Glucose	1.3 g	3 g
Distilled H_2O	1 liter	1 liter

9. Phage Mu Plates

Bactotryptone (Difco)	10 g
Yeast extract (Difco)	5 g
NaCl	10 g
Agar	10 g
Distilled H_2O	1 liter

Make the plates 1 mM $CaCl_2$ and 2.5 mM $MgSO_4$ by addition of sterile salts after autoclaving. For top agar, reduce the agar concentration by one-half.

10. Tetrazolium Plates

These are the most convenient plates to use for the isolation of carbohydrate-negative strains. If the cells grow on the added sugar, then their colonies first are white, and over several days turn pink and eventually faint red. The cells that do not use the sugar produce colonies that begin as pink, turn bright red after about 18 h, and remain red. The color indications on these plates are stable upon prolonged storage. Excessive crowding of the colonies inhibits the production of the red dye by sugar nonutilizers. Such inhibition becomes a problem at densities of about 500–1000 colonies per plate.

a. Combine 50 mg of 2,3,5-triphenyltetrazolium chloride (Sigma or Kodak) and 1 liter of distilled H_2O.
b. Dissolve all tetrazolium.
c. Add 25.5 g of Bacto antibiotic medium 2 (Difco).
d. Autoclave and cool.
e. Add 50 ml 20% sugar.

11. EMB and EMBO Plates

On these plates carbohydrate nonutilizers form pink colonies and carbohydrate utilizers form colonies of a dark color. On these plates utilizers outgrow nonutilizers, thus allowing utilizers to be detected when 10^9 nonutilizers and 100 utilizers are spread on the same plate. After 2 days the 100 utilizer-positives will have grown through the confluent lawn of nonutilizer-negatives and can be purified.

EMBO plates are the same as EMB with the exception that they do not have any carbohydrates.

Bactotryptone (Difco)	10 g
Yeast extract (Difco)	1 g
Agar	15 g
K_2HPO_4	2 g
Distilled H_2O	1 liter

Autoclave, cool, and add 40 ml of 1% eosin yellow (Difco), 13 ml of 0.5% methylene blue (Difco), and 50 ml of 20% sugar.

12. Xgal (5-bromo-4-chloro-3-indolyl-β-D-galactoside) Plates

An insoluble blue dye is produced when Xgal is cleaved by β-galactosidase. This is a particularly sensitive indicator for low levels of the enzyme.

Shortly before use, dissolve Xgal at 20 mg/ml in *N,N*-dimethyl formamide. Add this to any desired plate medium just before pouring to make a concentration of 40 μg/ml.

13. Arabinose Fucose Minimal Plates

D-Fucose, an analog of L-arabinose, inhibits arabinose induction of the ara operon via *araC* protein. Mutants resistant to fucose are constitutive and can be selected on these plates.

Autoclave 250 ml A plus B minimal medium, cool, and add the following:

Vitamin B_1, 2.5 ml of 1 mg/ml
L-Arabinose, 1.25 ml of 20% (wt/vol)
D-Fucose, 2.5 ml of 20% (wt/vol)

14. Ribitol Plates

Ribitol is phosphorylated by the ribulokinase of the *ara* operon, and when phosphorylated inhibits growth. *AraC* mutants unable to induce the *ara* operon or *araB* mutants which do not have significant ribulokinase activity are able to grow on these plates. Generally *araC* mutants form the larger colonies, but after extended incubation mucoid growth of undesired cells overgrows everything else.

Sterile A, recipe number 4	4.5 ml
Agar for plates, recipe number 4	4.5 ml
20% glycerol	0.1 ml

Vitamin B_1, 1 mg/ml	0.1 ml
20% arabinose	0.5 ml
20% ribitol	2.0 ml

Makes 3 tiny plates, 35 × 10 mm.

15. Stab-Agar for Storing Strains

Agar (Difco)	6 g
Bactotryptone (Difco)	10 g
NaCl	8 g
Cysteine-HCl	20 mg
Distilled H_2O	1 liter

16. Phage Lambda Suspension Medium and Storage Buffer

Tris-HCl, pH 7.5	10 ml, 1M
$MgCl_2$	10 ml, 1 M
Distilled H_2O	961 ml

For storage buffer, include gelatin at 0.05%.

17. TM Medium for Nitrosoguanidine Mutagenesis

Maleic acid	0.508 g
Tris	0.605 g
Distilled H_2O	100 ml

Adjust pH to 6.0 with NaOH.

18. 50 × TE Buffer for DNA

The TE buffer is 10 mM Tris, 1 mM EDTA, pH 8.0. It is convenient to make the following 50× stock that will be pH 8.0 when diluted with distilled H_2O.

Tris-base	60.55 g
EDTA	14.61 g
Distilled H_2O	800 ml

Adjust to pH 8.2 with HCl and fill to 1 liter with distilled water.

19. 2 × Freezing Medium for Cells

K_2HPO_4	12.6 g
Na^+-citrate	0.9 g
$MgSO_4 \cdot 7\ H_2O$	0.18 g

$(NH_4)_2SO_4$	1.8 g
KH_2PO_4	3.6 g
Glycerol	88 g

Fill to 1 liter with distilled water and sterilize by autoclaving.

20. 20 × SSC Buffer for DNA

NaCl	175 g
Na^+-citrate	88 g
H_2O to make	1 liter

Concentrated HCl, approximately 1 ml, is added to give pH 7.0.

21. Formamide

The commercial preparations of this chemical usually contain salt impurities as well as the hydrolysis products, ammonium formate, NH_4^+, and formic acid. In order not to seriously affect nucleic acid hybridization experiments, the formamide usually should have a conductivity below 40 μS and a 50% aqueous solution of it should have a pH below 7.5. The formamide may be cleaned by stirring 1 liter with 10 g of Norite A (Fisher) and 50 g of mixed bed resin (AG 501-X8(D), 20–50 mesh; Bio-Rad) at 4 °C for 2 h. Filter twice through Whatman #1 filter paper to remove the resin. If this does not work, recrystallize the formamide at −2 to −3 °C. Store aliquots at −70 °C.

22. Carrier DNA

Add the appropriate amount of TE, pH 8, to dry commercial DNA (Calbiochem or Sigma) to achieve a concentration of 5–10 mg/ml. Sonicate 5–10 times for 30 sec at a power setting that gives cavitation but no foaming. Monitor the temperature between sonications and do not allow it to go above 30 °C. Chill the DNA solution on ice between each sonication. Shake the DNA solution at moderate speeds (setting 3–5) on a New Brunswick shaker bath (G76) at room temperature, until the DNA is in solution. This may take several days. Sonicate the DNA once again using the procedure described above. Then dialyse 4 times for 4 h at 4 °C, against at least a 20-fold excess of TE, pH 8. Many commercial preparations contain a large quantity of salt that can be removed by this dialysis step. Use absorbance to determine the purity and concentration. Store at 4 °C.

23. Distilled Water

Distilled water can be remarkably filthy. One should be aware that some localities treat water by adding compounds, reportedly aromatic amines. Whatever is added, some of it comes over in distillation. These impurities can be removed by passing the water through an ion exchanger (Research Model 1, Illinois Water Treatment Co., Rockford, Ill.).

There are several common practices that introduce contaminants after distillation. Tygon tubing should not be used to carry distilled water. It leaches compounds into the water. It leaches faster at higher temperatures and so becomes even more of a problem with freshly distilled water. Greased stopcocks are also a source of problems. We recommend Teflon stopcocks that do not require grease. For water of the highest quality we redistill house-distilled water, pass it through only glass lines, and pour it from a jug rather than using a valve.

24. Chocolate Chip Cookies

This is a useful nutrient medium for laboratory workers.

WARNING: Try this at home. Eating (or smoking) in laboratories where radioactive, carcinogenic, or other toxic substances are found is dangerous!

1 cup flour
6 tablespoons brown sugar
6 tablespoons sugar
½ teaspoon salt
½ teaspoon baking soda
½ cup margarine
1 teaspoon vanilla
½ cup chopped nuts
1 cup chocolate chips
¼ teaspoon water
1 egg

Combine dry ingredients. Beat in egg. Add vanilla, morsels, and nuts. Bake 8–12 min at 350 °F.

Appendix II

Useful Numbers

a. 1 μCi = 3.7×10^4 dps
= 2.2×10^6 dpm

Avogadro's number = 6×10^{23}

b. Concentrations of standard commercial reagents

HCl	11.6 M (36% HCl)
Nitric acid	16.4 M
H_2SO_4	17.8 M
H_3PO_4	14.7 M (ortho; 85%)
Glacial acetic acid	17.4 M
2-mercaptoethanol	15.6 M

c. Typical *E. coli*

10^{-12} g wet weight per cell
10^{-13} g protein per cell
2×10^{-13} g dry weight per cell
OD_{500} (Zeiss) of 1.0 is about 4×10^8 cells/ml
One molecule per *E. coli* cell = 1×10^{-9} M

d. DNA

Molecular weight of an average base pair is approximately 650 daltons
10^6 daltons of double-stranded DNA is approximately 0.5 μm long
10^6 daltons of double-stranded DNA has approximately 1.5 kb
3 nmol of bases weigh approximately 1 μg

e. Approximate genome sizes

Coliphage lambda: 48 kb (Szybalski and Szybalski, 1979)
E. coli: 4×10^3 kb (Klotz and Zimm, 1972)

Drosophila melanogaster: 1.6 × 10^5 kb (Rasch et al., 1971)
Cow: 3 × 10^6 kb (Britten and Kohne, 1968)

f. Lengths of useful DNA-size standards

pBR322[a]				λ_{papa} CI_{857}[b]
Hae III (bp)	Hinf I (bp)	EcoR II (bp)	EcoR I (bp)	Hind III (kb)
587	1631	1857	4361	23.3
540	517	1060		9.31
504	506	928		6.46
458	396	383		4.26
434	344	121		2.23
267	298	13		1.93
234	221			0.54
213	220			
192	154			
184	75			
124				
123				
104				
89				
80				
64				
57				
51				
21				
18				
11				
7				

bp = base pairs.

[a]Sutcliffe, 1978.
[b]Wellauer et al., 1974; Szybalski and Szybalski, 1979.

Bibliography

Adelberg, E., Mandel, M., and Chen, G. (1965): Optimal conditions for mutagenesis by *N*-Methyl-*N'*-Nitro-*N*-Nitrosoguanidine in *Escherichia coli* K-12. Biochem Biophys Res Commun 18:788–795.

Adhya, S., Sen, A., and Mitra, S. (1971): The role of gene S. In *The Bacteriophage Lambda* (A.D. Hershey, ed.). New York, Cold Spring Harbor Laboratory, pp. 743–746.

Alberts, B. (1967): Fractionation of nucleic acids by dextran-polyethylene glycol two-phase systems. Methods Enzymol 12:566–581.

Albertsson, P.A. (1960): In *Partition of Cell Particles and Macromolecules* (Alquist and Wiksel, eds.). New York, Wiley Interscience.

Alikhanian, S.I., Iljina, T.S., Kalaeva, E.S., Kameneva, S., and Scakhodolec, V.V. (1966): A genetic study of thymineless mutants of *E. coli* K-12. Genet Res 8:83–100.

Allen, F.S., Gray, D.M., Roberts, G.P., and Tinoco, I., Jr. (1972): The ultraviolet circular dichroism of some natural DNAs and an analysis of the spectra for sequence information. Biopolymers 11:853–879.

Alwine, J.C., Kemp, D.J., and Stark, G.R. (1977): Method for detection of specific RNAs in agarose gels by transfer to diazobenzyloxymethyl-paper and hybridization with DNA probes. Proc Natl Acad Sci USA 74:5350–5354.

Anderson, E.H. (1946): Growth requirement of virus-resistant mutants of *Escherichia coli* strain "B." Proc Natl Acad Sci USA 32:120–128.

Aviv, J. and Leder, P. (1972): Purification of biologically active globin mRNA by chromatography on oligothymidylic acid cellulose. Proc Natl Acad Sci USA 69:1408–1412.

Bachmann, B. (1972): Pedigrees of some mutant strains of *Escherichia coli* K-12. Bacteriol Rev 36:525–557.

Bachmann, B.J. and Low, K.B. (1980): Linkage map of *Escherichia coli* K-12, edition 6. Microbiol Rev 44:1–56.

Barnes, W.M. (1977): Plasmid detection and sizing in single colony lysates. Science 195:393–394.

Barnett, T., Pachl, C., Gergen, J.P., and Wensink, P.C. (1980): The isolation and characterization of Drosophila yolk protein genes. *Cell* 21:729–738.

Benton, W.D. and Davis, R.W. (1977): Screening λgt recombinant clones by hybridization to single plaques in situ. Science 196:180–182.

Berger, S.L. (1975): Diethylpyrocarbonate: An examination of its properties in buffered solutions with a new assay technique. Anal Biochem 67:428–437.

Bochner, B., Huang, H., Schieven, G., and Ames, B. (1980): Positive selection for loss of tetracycline resistance. J Bacteriol 143:926–933.

Bolivar, F. (1978): Construction and characterization of new cloning vehicles. III. Derivatives of plasmid pBR322 carrying unique EcoRI sites for selection on EcoRI-generated recombinant DNA molecules. Gene 4:121–136.

Bolivar, F., Rodrigues, R.L., Betlach, M.C., and Boyer, H.W. (1977): Construction and characterization of new cloning vehicles. I. Ampicillin-resistant derivatives of the plasmid pMB9. Gene 2:75–93

Bonner, T.I., Brenner, D.J., Neufeld, B.R., and Britten, R.J. (1973): Reduction in the rate of DNA reassociation by sequence divergence. J Mol Biol 81:123–135.

Britten, R.J., Graham, D.E., and Neufeld, B.R. (1974): Analysis of repeating DNA sequences by reassociation. Methods Enzymol 29:373–418.

Britten, R.J. and Kohne, D.E. (1968): Repeated sequences in DNA. Science 161:529–541.

Burgess, R. (1969): A new method for the large-scale purification of *E. coli* DNA-dependent RNA polymerase. J Biol Chem 244:6160–6167.

Burgess, R. and Jendrisak, J. (1975): A procedure for the rapid, large-scale purification of *Escherichia coli* DNA-dependent RNA-cellulose chromatography. Biochemistry 14:4634–4638.

Burgess, R.R. and Travers, A.A. (1971): DNA-dependent RNA polymerase (EC 2.7.7.6). Proc Nucleic Acid Res 2:851–863.

Casadaban, M., Chou, J., and Cohen, S. (1980): In vitro gene fusions that join an enzymatically active beta-galactosidase segment to amino-terminal fragments of exogenous proteins: *E. coli* plasmid vectors for the detection and cloning of translational initiation signals. J Bacteriol 143:971–980.

Casadaban, M. and Cohen, S. (1979): Lactose genes fused to exogenous promoters in one step using a Mu-*lac* bacteriophage: in vivo probe for transcriptional control sequences. Proc Natl Acad Sci USA 76:4530–4533.

Cerdá-Olmedo, E., Hanawalt, P., and Guerola, N. (1968): Mutagenesis of the replication point by nitrosoguanidine. J Mol Biol 33:705–719.

Clark, A. and Margulies, A. (1965): Isolation and characterization of recombination-deficient mutants of *E. coli* K-12. Proc Natl Acad Sci USA 53:451–459.

Clewell, D.B. (1972): Nature of col E_1 plasmid replication in *Escherichia coli* in the presence of chloramphenicol. J Bacteriol 110:667–676.

Clewell, D.B. and Helinski, D.R. (1969): Supercoiled circular DNA-protein complex in *Escherichia coli:* Purification and induced conversion to an open circular DNA form. Proc Natl Acad Sci USA 62:1159–1166.

Cox, G.B., Luke R.K., Newton, N.A., O'Brien, I.G., and Rosenberg, H. (1970): Mutations affecting iron transport in *Escherichia coli.* J Bacteriol 104:219–226.

Craven, G.R., Steers, E., Jr., and Anfinsen, C.B. (1965): Purification, composition, and molecular weight of the β-galactosidase of *Escherichia coli* K-12. J Biol Chem 240:2468–2477.

Crombrugghe, B. de, Perlman, R., Varmus, H., and Pastan, I. (1969): Regulation of inducible enzyme synthesis in *E. coli* by cyclic adenosine 3′-5′ monophosphate. J Biol Chem 244:5828–5835.

Dagert, M. and Ehrlich, S.D. (1979): Prolonged incubation in calcium chloride improves the competence of *Escherichia coli* cells. Gene 6:23–28.

Dambly, C., Couturier, M., and Thomas, R. (1968): Control of development in temperate bacteriophages, II. Control of lysozyme synthesis. J Mol Biol 32:67–81.

Dugaiczyk, A., Boyer, H.W., and Goodman, H.M. (1975): Ligation of Eco RI endonuclease-generated DNA fragments into linear and circular structures. J Mol Biol 96:171–181.

Davidson, N. and Szybalski, W. (1971): *The Bacteriophage Lambda* (A.D. Hershey, ed.). New York, Cold Spring Harbor Laboratory.

Davies, J., Gilbert, W., and Gorini, L. (1964): Streptomycin, suppression, and the code. Proc Natl Acad Sci USA 51:883–888.

Davis, R.W. and Parkinson, J.S. (1971): Deletion mutants of bacteriophage lambda. J Mol Biol 56:403–423.

Demerec, M., Adelberg, E., Clark, A., and Hartman, P. (1966): A proposal for a uniform nomenclature in bacterial genetics. Genetics 54:61–76.

Denhardt, D.T. (1966): A membrane filter technique for the detection of complementary DNA. Biochem Biophys Res Commun 23:641–646.

Dische, Z. and Borenfreund, E. (1951): A new spectrophotometric method for the detection and determination of keto sugars and trioses. J Biol Chem 192:583–587.

Doub, L. and Vandenbelt, J. (1949): The ultraviolet absorption spectra of simple unsaturated compounds. J Am Chem Soc 71:2414–2420.

Dove, W.F. and Davidson, N. (1962): Cation effects on the denaturation of DNA. J Mol Biol 5:467–478.

Edlin, G. and Broda, P. (1968): Physiology and genetics of the "RNA control" locus in *E. coli.* Bacteriol Rev 32:206–226.

Eilen, E., Pampeno, C., and Krakow, J. (1978): Production and properties of the α core derived from cyclic AMP receptor protein of *Escherichia coli.* Biochemistry 17:2469–2473.

Englesberg, E. (1966): Isolation of mutants in the L-arabinose gene-enzyme complex. Methods Enzymol. 9:15–21.

Epstein, W., Rothman-Denes, L., and Hesse, J. (1975): Adenosine 3′-5′-cyclic monophosphate as mediator of catabolite repression in *E. coli.* Proc Natl Acad Sci USA 72:2300–2304.

Felsenfeld, G. (1971): Analysis of temperature-dependent absorption spectra of nucleic acids. Proc Nucleic Acid Res. 2:233–244.

Fiandt, M., Honigman, A., Rosenvold, E.C., and Szybalski, W. (1977): Precise measurement of the b2 deletion in coliphage lambda. Gene 2:289–293.

Fiil, N. (1969): A functional analysis of the rel gene in *E. coli.* J Mol Biol 45:195–203.

Frost, G. and Rosenberg, H. (1975): Relationship between the *ton*B locus and iron transport in *E. coli.* J Bacteriol 124:704–712.

Gellert, M. (1971): DNA ligase from *E. coli.* Proc Nucleic Acid Res 2:875–888.

Gergen, J.P., Stern, R.H., and Wensink, P.C. (1979): Filter replicas and permanent collections of recombinant DNA plasmids. Nucleic Acids Res 7:2115–2136.

Gilbert, W. and Müller-Hill, B. (1966): Isolation of the *lac* repressor. Proc Natl Acad Sci USA 56:1891–1898.

Goldberg, A. and Howe, M. (1969): New mutations in the S cistron of bacteriophage lambda affecting host cell lysis. Virology 38:200–202.

Good, N.E., Winget, G.D., Winter, W., Connelly, T.N., Izawa, S., and Singh, R.M.M. (1966): Hydrogen ion buffers for biological research. Biochemistry 5:467–477.

Gorini, L. and Kaufman, H. (1960): Selecting bacterial mutants by the penicillin method. Science 131:604–605.

Greenblatt, J. (1972): Positive control of endolysin synthesis in vitro by the gene N protein of phage lambda. Proc Natl Acad Sci USA 69:3606–3610.

Greenblatt, J. and Schleif, R. (1971): Arabinose C protein: Regulation of the arabinose operon in vitro. Nature New Biol 233:166–170.

Grossman, L., and Moldave, K. (1980): Nucleic Acids, Part 1. Methods Enzymol 65.

Grunstein, M. and Hogness, D.S. (1975): Colony hybridization: A method for the isolation of cloned DNAs that contain a specific gene. Proc Natl Acad Sci USA 72:3961–3965.

Guardiola, J., De Felice, M., Klopotowski, R., and Iaccarino, M. (1974): Mutations affecting the different transport systems for isoleucine, leucine, and valine in *Escherichia coli* K-12. J Bacteriol 117:393–405.

Guarente, L., Lauer, G., Roberts, T., and Ptashne, M. (1980): Improved methods for maximizing expression of a cloned gene: A bacterium that synthesizes rabbit beta-globin. Cell 20:543–553.

Haseltine, W.A. and Baltimore, D. (1976): Size of murine RNA tumor virus-specific nuclear RNA molecules. J Virol 19:331–337.

Helling, R.B., Goodman, H.M., and Boyer, H.W. (1974): Analysis of endonuclease R.Eco RI fragments of DNA from lamboid bacteriophages and other viruses by agarose gel electrophoresis. J Virol 14:1235–1244.

Hershey, A.D. (ed.) (1971): *The Bacteriophage Lambda.* Cold Spring Harbor Laboratory.

Heyneker, H.L., Shine, J., Goodman, H.J., Boyer, H., Rosenberg, J., Dickerson, R.E., Narang, D.A., and Riggs, A.D. (1976): Synthetic lac operator DNA is functional in vivo. Nature 263:748–752.

Hirota, Y. (1960): The effect of acridine dyes on mating type factors in *Escherichia coli.* Proc Natl Acad Sci USA 46:57–64.

Hirsh, J. and Schleif, R. (1976): High resolution electron microscopic studies of genetic regulation. J Mol Biol 108:471–490.

Hogness, D.S. and Simmons, J.R. (1964): Breakage of λdg DNA: Chemical and genetic characterization of each isolated half-molecule. J Mol Biol 9:411–438.

Hudson, B., Dawson, J.H., Desiderio, R., and Mosher, C.W. (1977): Ethidium analogs with improved resolution in the dye-buoyant density procedure. Nucleic Acids Res. 4:1349–1359.

Hutton, J.A. (1977): Renaturation kinetics and thermostability of DNA in aqueous solutions of formamide and urea. Nucleic Acids Res 4:3537–3555.

Inuzuka, N., Nakamura, S., Inuzuka, M., and Tomoeda, M. (1969): Specific action of sodium dodecyl sulfate on the sex factor of *E. coli* K-12 and Hfr strains. J Bacteriol 101:827–839.

Kacian, D.L., Watson, K.F., Burny, A., and Spiegelman, S. (1971): Purification of the DNA polymerase of avian myeloblastosis virus. Biochem Biophys Acta 246:365–383.

Kaiser, D. and Masuda, T. (1973): In vitro assembly of bacteriophage lambda heads. Proc Natl Acad Sci USA 70:260–264.

Kelly, R.B., Cozzarelli, N.R., Deutscher, M.P., Lehman, I.R., and Kornberg, A. (1970): Enzymatic synthesis of deoxyribonucleic acid. XXXII. Replication of duplex deoxyribonucleic acid by polymerase at a single strand break. J Biol Chem 245:39–45.

Kirby, K.S. (1957): A new method for the isolation of deoxyribonucleic acids: evidence on the nature of bonds between deoxyribonucleic acid and protein. Biochem J 66:495–504.

Kleckner, N., Roth, J., and Botstein, D. (1977): Genetic engineering in vivo using translocatable drug-resistance elements. J Mol Biol 116:115–159.

Klotz, L.C. and Zimm, B.H. (1972): Size of DNA determined by viscoelastic measurements: Results on bacteriophages, *Bacillus subtilis* and *Escherichia coli.* J Mol Biol 72:779–800.

Kushner, S.R. (1978): Improved methods for transformation of *Escherichia coli* with col E1 derived plasmids. In *International Symposium on Genetic Engineering: Scientific Developments and Practical Application. Milan, Italy. March 29–31, 1978. Giovanni Lorenzini Foundation. Vol 2* (H.W. Boyer and S. Micosia, eds.). Amsterdam, North-Holland. pp. 17–23.

Laemmli, U.K. (1970): Cleavage of structural proteins during the assembly of the head of bacteriophage T4. Nature 227: 680–685.

Laird, C.D., McConaughy, B.L., and McCarthy, B.J. (1969): Rate of fixation of nucleotide substitutions in evolution. Nature 224:149–154.

Lennox, E. (1955): Transduction of linked genetic characters of the host by bacteriophage P1. Virology 1:190–206.

Lis, J.T. and Schleif, R. (1975): Size fractionation of double-stranded DNA by precipitation with polyethylene glycol. Nucleic Acids Res 2:383–389.

Lis, J.T., Prestidge, L., and Hogness, D.S. (1978): A novel arrangement of tandemly repeated genes at a major heat shock site in *D. melanogaster.* Cell 14:901–919.

Little, J.W., Lehman, I.R., and Kaiser, A.D. (1967): An exonuclease induced by bacteriophage λ. I. Preparation of the crystalline enzyme. J Biol Chem 242:672–678.

Low, B. (1973): Rapid mapping of conditional and auxotrophic mutations in *E. coli* K-12. J Bacteriol 113:798–812.

Low, K.B. (1972): *Escherichia coli* K-12 F-prime factor, old and new. Bacteriol Rev 36:587–607.

McConaughy, B.L., Laird, C.B., and McCarthy, B.J. (1969): Nucleic acid reassociation in formamide. Biochemistry 8:3289–3295.

McDonell, M.S., Simon, M.N., and Studier, F.W. (1977): Analysis of restriction fragments of T7 DNA and determination of molecular weights by electrophoresis in neutral and alkaline gels. J Mol Biol 110:119–146.

McMaster, G.K. and Carmichael, G.G. (1977): Analysis of single- and double-stranded nucleic acids on polyacrylamide and agarose gels by using glyoxal and acridine orange. Proc Natl Acad Sci USA 74:4835–4838.

Maaloe, O. and Kjeldgaard, N. (1966): *Control of Macromolecular Synthesis.* New York, Benjamin.

Mandel, M. and Higa, A. (1970): Calcium-dependent bacteriophage DNA infection. J Mol Biol 53:159–162.

Maniatis, T., Hardison, R.C., Lacy, E., Lauer, J., O'Connell, C., and Quon, D. (1978): The isolation of structural genes from libraries of eukaryotic DNA. Cell 15:687–701.

Maniatis, T., Jeffrey, A., and van deSande, H. (1975): Chain length determination of small double- and single-stranded DNA molecules by polyacrylamide gel electrophoresis. Biochemistry 14:3787–3794.

Marmur, J. and Doty, P. (1962): Determination of the base composition of deoxyribonucleic acid from its thermal denaturation temperature. J Mol Biol 5:109–118.

Martinson, H.G. (1973): The nucleic acid-hydroxyapatite interaction. II. Phase transitions in the deoxyribonucleic acid-hydroxyapatite system. Biochemistry 12:145–150.

Maxam, A.M. and Gilbert, W. (1977): A new method for sequencing DNA. Proc Natl Acad Sci USA 74:560–564.

Miller, J. (1972): *Experiments in Molecular Genetics.* New York, Cold Spring Harbor Laboratory.

Neidhardt, F.C., Bloch, P.L., and Smith, D.F. (1974): Culture medium for enterobacteria. J Bacteriol 119:736–747.

Oeschger, M. and Berlyn, M. (1974): A simple procedure for localized mutagenesis using nitrosoguanidine. Mol Gen Genet 134:77–83.

O'Farrell, P.H. (1975): High resolution two-dimensional electrophoresis of proteins. J Biol Chem 250:4007–4021.

O'Farrell, P.H., Kutter, E., and Nakanishi, M. (1980): A restriction map of the bacteriophage T4 genome. Mol Gen Genet 179:421–435.

O'Farrell, P.Z., Goodman, H.M., and O'Farrell, P.H. (1977): High resolution two-dimensional electrophoresis of basic as well as acidic proteins. Cell 12:1133–1142.

Parkinson, J. and Huskey, R. (1971): Deletion mutants of bacteriophage lambda. J Mol Biol 56:369–384.

Parks, J., Gottesman, M., Perlman, R., and Pastan, I. (1971): Regulation of galactokinase synthesis by cyclic adenosine 3′,5′-monophosphate in cell-free extracts of *E. coli.* J Biol Chem 246:2419–2424.

Paterson, B.M., Marciani, D.J., and Papas, T.S. (1977): Cell-free synthesis of the precursor polypeptide for avian myeloblastosis virus DNA polymerase. Proc Natl Acad Sci USA 74:4951–4954.

Patrick, J. and Lee, N. (1968): Purification and properties of an L-arabinose isomerase from *E. coli.* J Biol Chem 243:4312–4318.

Pelham, H.R.B. and Jackson, R.J. (1976): An efficient RNA-dependent translation system from reticulocyte lysates. Eur J Biochem 67:247–251.

Polisky, B., Greene, P., Garfin, D.E., McCarthy, B.J., Goodman, H.M., and Boyer, H.W. (1975): Specificity of substrate recognition by the *Eco*RI restriction endonuclease. Proc Natl Acad Sci USA 72:3310–3314.

Ptashne, M. (1971): Repressor and its action. In *The Bacteriophage Lambda* (A.D. Hershey, ed.). New York, Cold Spring Harbor Laboratory, pp. 221–237.

Ptashne, M., Bachmann, K., Humayun, M., Jeffrey, A., Maurer, R., Meyer, B., and Aauer, R. (1976): Autoregulation and function of a repressor in bacteriophage lambda. Science 194:156–161.

Radloff, R., Bauer, W., and Vinograd, J. (1967): A dye-buoyant-density method for the detection and isolation of closed circular duplex DNA: The closed circular DNA in Hela cells. Proc Natl Acad Sci USA 57:1514–1521.

Rasch, E.M., Barr, H.J., and Rasch, R.W. (1971): The DNA content of sperm of *Drosophila melanogaster.* Chromosoma 33:1–18.

Rave, N., Crkyenjakov, R., and Boedtker, H. (1979): Identification of procollagen mRNAs transferred to diazobenzyloxymethyl paper from formaldehyde agarose gels. Nucleic Acids Res 6:3559–3567.

Richardson, C.C. (1971): Polynucleotide kinase from *Escherichia coli* infected with bacteriophage T4. Proc Nucleic Acid Res 2:815–828.

Rigby, P.W.J., Dieckmann, M., Rhodes, C., and Berg, P. (1977): Labelling of deoxyribonucleic acid to high specific activity in vitro by nick translation with DNA polymerase I. J Mol Biol 113:237–251.

Roberts, B.E. and Paterson, B.M. (1973): Efficient translation of tobacco mosaic virus RNA and rabbit globin 9S RNA in a cell-free system from commercial wheat germ. Proc Natl Acad Sci USA 70:2330–2334.

Rosenvold, E.C. and Honigman, A. (1977): Mapping of Ava I and Xma I cleavage sites in bacteriophage

DNA including a new technique of DNA digestion in agarose gels. Gene 2:273-288.

Roychoudhury, R., Jay, E., and Wu, R. (1976): Terminal labelling and addition of homopolymer tracts to duplex DNA fragments by terminal deoxynucleotidyl transferase. Nucleic Acids Res 3:863-877.

Sato, H. and Miura, K.I. (1963): Preparation of transforming deoxyribonucleic acid by phenol treatment. Biochim Biophys Acta 72:619-629.

Sax, N.I. (1979): Dangerous Properties of Industrial Chemicals. New York, Van Nostrand Reinhold.

Schechtman, M., Algre, J., and Roberts, J. (1980): Assay and characterization of late gene regulators of bacteriophage 82 and lambda. J Mol Biol 142:269-288.

Schleif, R. (1967): Control of production of ribosomal protein. J Mol Biol 27:41-55.

Schleif, R. (1972): Fine-structure deletion map of the *Escherichia coli* L-arabinose operon. Proc Natl Acad Sci USA 69:3479-3484.

Schleif, R. and Lis, J.T. (1975): The regulatory region of the L-arabinose operon: Genetic and physiological study. J Mol Biol 95:417-431.

Schleif, R., Greenblatt, J., and Davis, R. (1971): Dual control of arabinose genes of transducing phage lambda-*dara.* J Mol Biol 59:127-150.

Schleif, R., Hess, W., Finkelstein, S., and Ellis, D. (1973): Induction kinetics of the L-arabinose operon of *E. coli.* J Bacteriol 115:9-14.

Schultz, G. and Schirmer, R. (1978): Principles of Protein Structure. New York, Academic Press, pp. 149-165.

Schwartz, M. (1967): Sur l'existence chez *Escherichia coli* K-12 d'une regulation commune à la biosynthése des récetteurs du bactériophage lambda et au métabolisme du maltose. Ann Inst Pasteur 113:685-704.

Seeburg, P.H., Shine, J., Martial, J.A., Baxter, J.D., and Goodman, H.M. (1977): Nucleotide sequence and amplification in bacteria of structural gene for rat growth hormone. Nature 270:486-494.

Sgaramella, V., van de Sande, J.H., and Khonana, H.G. (1970): Studies on polynucleotides, C. A novel joining reaction catalyzed by T4-polynucleotide ligase. Proc Natl Acad Sci USA 67:1468-1475.

Signer, E. (1971): General recombination. In *The Bacteriophage Lambda* (A.D. Hershey, ed.). Cold Spring Harbor Laboratory.

Signer, E.R. (1970): On control of lysogeny in phage λ. *Virology 40:624-633.*

Smith, H.O. and Birnstiel, M.L. (1976): A simple method for DNA restriction site mapping. Nucleic Acids Res 3:2387-2398.

Southern, E.M. (1975): Detection of specific sequences among DNA fragments separated by gel electrophoresis. J Mol Biol 98:503-517.

Standler, S. and Adelberg, E.A. (1972): Temperature dependence of sex factor maintenance in *E. coli* K-12. J Bacteriol 109:447-449.

Stanier, R.Y. and Adelberg, E.A. (1970): *The Microbial World,* 3rd ed. Englewood Cliffs, N.J., Prentice-Hall.

Stent, G. (1963): *Molecular Biology of Bacterial Viruses.* San Francisco, Freeman.

Stent, G. and Brenner, S. (1961): A genetic locus for the regulation of RNA synthesis. Proc Natl Acad Sci USA 47:2005-2014.

Storti, R.V., Coen, D.M., and Rich, A. (1976): Tissue-specific forms of actin in the developing chick. Cell 8:521-527.

Studier, F. (1969): Effects of the conformation of single-stranded DNA on renaturation and aggregation. J Mol Biol 41:199-209.

Sussman, R. and Jacob, F. (1962): Sur un systeme de répression thermosensible chez le bactériophage lambda d'*Escherichia coli.* C R Acad Sci 254:1517-1519.

Sutcliffe, J.G. (1978): pBR322 restriction map derived from the DNA sequence: Accurate DNA size markers up to 4361 nucleotide pairs long. Nucleic Acids Res 5:2721-2728.

Swanstrom, R. and Shenk, P. (1978): X-ray intensifying screens greatly enhance the detection by autoradiography of the radioactive isotopes ^{32}P and ^{125}I. Anal Biochem 86:184-192.

Szybalski, E.H. and Szybalski, W. (1979): A comprehensive molecular map of bacteriophage lambda. Gene 7:217-270.

Tartof, K.D. and Perry, R.P. (1970): The 5S genes of *Drosophila melanogaster.* J Mol Biol 51:171-183.

Thomas, M., Cameron, J., and Davis, R. (1974): Viable molecular hybrids of bacteriophage lambda and eukaryotic DNA. Proc Natl Acad Sci USA 71:4579-4583.

Thomas, P.S. (1980): Hybridization of denatured RNA and small DNA fragments transferred to nitrocellulose. Proc Natl Acad Sci USA 77:5201-5205.

Ullrich, A., Shine, J., Chirgwin, J., Pictet, R., Tischer, E., Rutter, W.J., and Goodman, H.M. (1977): Rat insulin genes: Construction of plasmids containing the coding sequences. Science 196:1313-1319.

Wahl, G.M., Stern, M., and Stark, G.R. (1979): Efficient transfer of large DNA fragments from agarose gels to diazobenzyloxymethyl-paper and rapid hybridization by using dextran sulfate. Proc Natl Acad Sci USA 76:3683-3687.

Wang, C.C. and Newton, A. (1969): Iron transport in *Escherichia coli*: Relationship between chromium sensitivity and high iron requirement in mutants of *Escherichia coli.* J Bacteriol 98:1135-1141.

Weber, K. and Osborn, M. (1969): The reliability of molecular weight determinations by dodecyl sulfate-polyacrylamide gel electrophoresis. J Biol Chem 244:4406-4412.

Weiss, B., Jacquemin-Sablon, A., Live, T.R., Fareed, G.C., and Richardson, C.C. (1968): Enzymatic breakage and joining of deoxyribonucleic acid. VI. Further purification and properties of polynucleotide ligase from *Escherichia coli* infected with bacteriophage T4. J Biol Chem 243:4543-4555.

Wellauer, K., Reeder, R.H., Carroll, D., Brown, D.D., Deutch, A., Higashinakayawa, T., and Dawid, I.B. (1974): Amplified ribosomal DNA from *Xenopus laevis* has heterogeneous spacer lengths. Proc Natl Acad Sci USA 71:2823-2827.

Wensink, P.C., Finnegan, D.J., Donelson, J.E., and Hogness, D.S. (1974): A system for mapping DNA sequences in the chromosomes of *Drosophila melanogaster*. Cell 3:315-325.

Wensink, P.C., Tabata, S., and Pachl, C. (1979): The clustered and scrambled arrangement of moderately repetitive elements in *Drosophila* DNA. Cell 18:1231-1246.

Wetmur, J.G. and Davidson, N. (1968): Kinetics of renaturation of DNA. J Mol Biol 31:349-370.

Wheeler, E.L. (1958): *Scientific Glassblowing*. New York, Interscience.

Wiegand, R.C., Godson, G.N., and Radding, C.M. (1975): Specificity of the SI nuclease from *Aspergillus oryzae*. J Biol Chem 250:8848-8855.

Yates, J., Arfstein, A., and Nomura, M. (1980): In vitro expression of *E. coli* ribosomal protein genes: Autogenous inhibition of translation. Proc Natl Acad Sci USA 77:1837-1841.

Zubay, G., Chambers, D.A., and Cheong, L.C. (1970): Cell-free studies on the regulation of the *lac* operon. In *The Lactose Operon*. (J.R. Beckwith and D. Zipser, eds.). New York, Cold Spring Harbor Press, pp. 375-391.

Index

A_{260}: A_{280} ratio
  after Dextran–PEG phase partition, 67
  of DNA, 90
Absorbance
  DNA, 89
  nitrophenyl from ONPG, 45
  protein, 74
  RNA, 89–90
  triphosphates, 111
Acridine orange, 19
Acrylamide
  extracting DNA from, 122
  protein gels, 78–88
  DNA gels, 114–125
Aeration of cells, 4, 6, 11–13
Agarose
  DNA and RNA gels, 114–125
  electrophoresis of double- and single-stranded DNA in the same gel, 119
  extracting DNA from, 123
  gel aparatus, 115
Airborne phage contamination, 5–6, 31
Alkaline hydrolysis of RNA, 143
Alkaline phosphatase
  assay, 44
  digestion of DNA, 144
Alumina, 15
Amino acid starvation of bacteria, 4
Ammonium sulfate
  precipitation of protein, 62–64, 78
  to concentrate proteins, 76
  to remove PEG, 66
Antifoam, 9, 11
Arabinose fucose minimal plates, 200
Arabinose isomerase
  assays, 46–50
  units, 50
Assays
  arabinose isomerase, 46–50
  β-galactosidase, 43–45
  conductivity, 72–73
  coupled transcription–translation, 56–60
  lysozyme, 50–52
  nucleic acid, 89–91
  protein, 67, 74–76
  ribulokinase, 52–56
  RNA polymerase, 45–46
Autoradiography
  DNA and RNA gels, 114
  general, 184
  protein gels, 87–88
  screening phage plaques, 147–148
  screening colonies, 149–151
  Southern transfers, 152–156
Avian myeloblastosis virus reverse transcriptase, 141
Autoclaving, 195–196

β-galactosidase
  assay, 43–45
  molecular weight, 45
  molecules/cell, 45
  units/mg, 45
  turnover number, 44
Background in filter assays
  RNA polymerase, 46
  ribulokinase, 54
  TCA precipitation, 46, 96–97
Background in nucleic acid hybridization filter assays
  non-specific DNA binding, 153–156
  non-specific hybridization, 145
  RNase digestion to lower background, 150

Bacteria, 1–26
Bacterial alkaline phosphatase, 44, 144, 145
BAP, 44, 144, 145
Beckman ultracentrifuges and rotors, 178–179
Biuret protein assay, 74
Blotting DNA and RNA, 152–156
Blunt ends, 129
  making by S1 digestion, 130
Bovine serum albumin, 122
Bromphenol, 82, 118
Buffer gels, 79
Buffers, 177–178

C600, 4, 31
Cacodylate buffers, 178
Carboys, use for growing cells, 11–16
Carcinogens, 103, 191
Carrier DNA, 202
Carrier tRNA, 96, 97
Catabolite repression, 44
Cautions about laboratory hazards, 190–194
cDNA, 141
Cell
  density, 9–11
  health indication by streaming birefringence, 5
  lysis by phage lambda, 28
  lysis for plasmid DNA extraction, 101, 103
Centrifuge, 13, 178–179
Chemical resistances of plastics, 174
Chilling cells, large-scale, 13
Chloramphenicol, purity, 105
Chromatography
  ascending aqueous, 54–56
  buffers, 70
  proteins and nucleic acids, 68–73
  triphosphates, 113
Chocolate chip cookies, 203
Chromic ion effect on cell growth, 4
CI_{857}, 27
Cleaning DNA, 93–96
Colony hybridization, 146–151, 184
Colony screening, 149–151
Columns, 68–73
Columns, going dry, prevention, 68
Concentrating protein solutions, 76–77
Concentrations of standard commercial reagents, 205
Conductivity, in chromatography, 72
Coomassie Brilliant Blue, 87
Cotransduction, 22
Cross-links between DNA strands, 187
CsCl
  block gradient, 33
  density gradients for DNA banding, 110
  purifying, 106
  recovering, 188
  refractive index, 33, 103
Curing F-factors, 19
Cycloserine selection, 19, 136
Cysteine–carbazole test, 47

DBM paper, 157–158
DEAE, 52, 69
DEAE–cellulose, 52, 69–73, 93
DEAE filter paper, 52, 53, 55
DeLaval gyrotester, 13–14
Deletions, phage lambda, 40
  detection, 43
Denaturing proteins, 61
Denhardt's solution, 154
Density mutants of lambda phage, 35
DEPC
  treatment of solutions, 108
  sterilization of solutions, 170–171
  treatment of plastic and glassware, 168
Dephosphorylation of 5′ ends, 44, 144
Deprotenizing DNA, 93–96, 97–103
Deprotenizing RNA, 168
Desiccators, 190
Dextran, 64
Dialysis tubing, properties of, 186
Diazobenzyloxymethyl paper, 157
Dichromate in concentrated sulfuric acid, 175
Diethyl pyrocarbonate
  sterilization of solutions, 108, 170–171
  treatment of plastic and glassware, 168
Dimethylsulfoxide
  hazard, 192
  solvent, 88
  treatment to disaggregate RNA, 170–171
Diphenyloxazole, 88
Distilled water, 203
Distilling phenol, 187–189
DMSO, 88
  treatment to disaggregate RNA, 170
DNA
  absorbance, 89
  BAP treatment, 144
  binding to nitrocellulose paper, 154
  binding to DBM paper, 158
  chromatography with hydroxyapatite, 93–94
  chromatography with DEAE–cellulose, 93–94
  cleaning, 93–94
  complementary to RNA, 141
  controlled hydrolysis in gels, 154
  damage from UV light, 122

daltons/average base pair, 205
dissolving, 99
electroelution from gels, 124
electrophoresis, staining DNA in gels, 121
electrophoresis, 114–125
extracting from gels, 122
filter blots, 152
fluorescence, 90
from embryos, 110
gel electrophoresis, 114–125
hybridization, 145
hybridization to filter-bound DNA, 156
hybridization while on DBM paper, 158
isolation, *Drosophila* DNA, 108–111
isolation, *E. coli* DNA, 98–99
isolation from single colony, 151–152
isolation from most higher organisms, 108
isolation, lambda DNA by SDS, 100
isolation, lambda DNA by phenol, 99
isolation, plasmid DNA from a single colony, 151
isolation, plasmid DNA, 101–106
isolation, plasmid DNA, large-scale, 106
isolation, supercoils, 102
kb/dalton, 205
kinasing, 143
labelling, 138
length/dalton, 205
length vs. migration velocity in gels, 114–124
ligating, 120, 129–132
measuring concentration, 89–92, 108
molecular length (weight) standards, 126, 206
molecular weights, 205
moles/μg, 205
nicking, 139
precipitation with ethanol, 95–96
precipitation with polyethylene glycol, 97–98
precipitation with TCA, 96–97
purity, 90, 111
rearrangements, detection, 43
removing carbohydrate from, 111
separating double-stranded from single-stranded, 95
shear breaking, 129–130
single-stranded, 119
Southern hybridization, 155
separation range for electrophoresis gels, 114
size fractionation with polyethylene glycol, 97
size standards, 126, 206
storage, 92
DNase stock solutions, 139
Double lysogens, 38
Drawing figures, 179–180
Drug-resistance elements, 26
dT columns, 169
dT tails, 133
Dye markers for DNA electrophoresis, 118

***E.** coli*
DNA polymerase I, 138
genetic nomenclature, 2
genetic markers, 21–25
genetics, 1–26
growth for experiments, 3–15
mutagenesis, 17–18
mutants, 2
opening cells, 15–16
radiolabelling, 16–17
RNA polymerase, 140
strains, 1–3
transformation with plasmid DNA, 134
Egg-white lysozyme, 16, 52, 103, 106
Electrical hazards, 192–193
Electroelution of DNA from gels, 124
Electrophoresis apparatus
horizontal, 116–118
vertical, 83
Eluting columns, 71–73
EMB–maltose plates, 37
EMB plates, 199
EMBO plates, 199
End-labelling of DNA or RNA, 143–145
Endopeptidase lambda, 50
Episomes, 8
curing, 19–20
Equilibrium banding of DNA, 101–105
Ethanol precipitation of DNA, 93, 95–96
Ethidium analogs for staining DNA and for density gradient separation of DNAs, 106
Ethidium bromide, 90, 121
CsCl density gradients, 101, 103
DNA interaction, 101
removal from DNA, 104, 106
stock solutions, 121
Excluded volume of columns, 77
Exponential gradient gels, 85
Extracting DNA from acrylamide and agarose gels, 122–125

F-factors, 19
large-scale transfers, 24
transfer, 8
Figures, drawing and lettering, 179–180
Film response, 181–183
Filter assays
nick translation, 138–140

Filter assays *(cont'd)*
  ribulokinase, 54
  RNA polymerase, 46
  TCA precipitation, 46, 96–97, 163
Fines, removal, 69
Fire hazards, 191
Fluorescence of ethidium bromide–DNA, 90
Fluorescent intensifying screen, 184
Fluorography, 87, 184
Folin-Ciocalteau reagent, 76
Formamide
  effect on T_m of nucleic acid, 14
  purification, 202
Fraction collectors, 68–69
Freezing medium for cells, 201
Fusaric acid selection of tet^s cells, 137

Galactokinase, 52
Gas dispersion stones, 11
Gel electrophoresis
  nucleic acids, 114–124
  proteins, 78–88
Gel filtration, 66, 69–73
Gel photography, 87, 121–122
Gene fusions, detection, 43
Genetic crosses
  large-scale, 24
  small-scale, 7–9
Genetic instability, 3
Genetic maps, 2
Genome sizes, 205
Glass beads, 15
Glass containers, 173
Glass electrodes, 176
Glyoxal agarose gels, 119
Good buffers, 178
Gradients
  gels, 79
  for columns, 71–73
  maker, 86
  sucrose, 170–172
Grinding cells, 15–16
Ground fault interrupters, 192–194
Growing lambda stocks, 39–31
  large-scale, 31–35

Hazards, laboratory, 190–194
HCl gas, 159
Heat-sealable plastic bags, 155
Hfr, 8, 21
High-temperature episome curing, 20
Homopolymer tail addition to DNA, 133
Homopolymer tails, 130
Horizontal gel electrophoresis, 116
Hybridization
  general aspects, 145–146
  hybrid selection and purification, 156–159
  screening colonies, 149–151
  sensitivity, 156
  Southern transfers, 152–156
  to filter-bound DNA, 155
Hydroxyapatite, 69, 93–94
Hyperchromicity, DNA, 90

Immunity to phage lambda, 40
In vitro packaging of λ phage DNA, 27, 147–148
In vitro radiolabelling of DNA and RNA, 137–145
In vitro translation systems from higher organisms, 161–166
Included volume, 77
Induction of lambda receptor site, 35
Intensifying screens, autoradiography, 185–186
Ion exchange chromatography, 69–73

Joining the ends of DNA molecules, 129, 130–132

Kinases, 52
Kinasing DNA, 143–145

Laboratory safety, 190–194
Laemmli gels, 81
Lambda, 27–41
  exonuclease, 132
  growth, large-scale, 31–33
  lysogen, 32–37
  mutants, density, 35
  phage, 27, 197
  phage purification, 33
  phage recombination, 35
  plate stocks, 29–31, 149
  purification, 33–35
  receptor, 35
  resistance-indicating plates, 37
  resistant mutants, 37
  sensitivity, bacterial, 38
  storage buffer, 201
  suspension medium, 201
  titering, 28–29
  vector in cloning, 147–149
  vir, 37
LB media and plates, 196
Lettering figures, 179–180
Ligating DNA, 129, 130–132, 136
Light-scattering by bacteria, 10

Liquid polymer phase partitioning, 64–68
Loading columns, 71
Loading rotors, 179
Lowry assay for dilute samples, 75
Lowry protein assay, 75
  for dilute samples, 75–76
  PEG effect on, 67
Luer fittings, 68
Lysing cells, 15–16, 103
Lysogenization of P1 transductants, 24
Lysogens, making, 38–40
Lysogeny, 22
Lysozyme, 16, 50, 106

M10 liquid minimal medium, 197
m-nitrobenzylpyridinium chloride, 157
Maltose uptake and lambda resistance, 37
Mapping restriction endonuclease sites on DNA, 125–127
Mating *E. coli,* 7, 8, 24
Micrococcal nuclease digestion of lysate, 164
Minimal medium for plates, 197
Mu plates, 199
Mutagenesis, 3
  nitrosoguanidine, 17–18
Mutant enrichment, 18–19
Mutant phage identification, 36–38

Nalidixic acid, 8, 21
NBM paper, 157–158
NBPC synthesis, 159
Negatives, photographic, 181–183
Nick translation, 138–140
Nitex cloth, 109
Nitrocellulose paper, 146, 153–156, 162
Nitrophenyl, 43–45
Nitrosoguanidine, 17–18, 192
Nitrosoguanidine mutagenesis, 3, 17–18, 201
Nonspecific binding of triphosphates to Millipore filters, 46, 54, 96–97
NTG, 17–18
Nuclease inactivation, 108, 168, 170
Nucleic acid
  concentration, measurement of, 89–92
  hybridization reactions, general aspects, 145
  hybridization, 146–148, 150, 155–156, 158–159
  hyperchromicity, 89
  oligomer, 96
  optical absorbance, 89
  precursor incorporation assay, 112
  purity, measurement of, 89
Nucleoside chromatography, 112
Nucleoside triphosphate solutions, preparation of, 111
Nylon Nitex cloth, 109

Oligo-(dT)-cellulose column, 169
ONPG, 43
Opening cells, 15–16, 103
Organic chemicals, 189, 191–192
Organic solvents, 191
Orthonitrophenyl-galactoside, 43
Osmotic shocking, 18

P1, plates, 198
  titration, 23
  top agar, 198
  transduction, 3, 23
p-nitrophenyl, 44–45
PEG, 64–68
  effects on enzymes, 67
  removal, 66
PEI strips, 113
Penicillin, 18–19
Permeablizing cells, 44, 48, 54
Petroff–Hauser cell chamber, 9–10
pH meters, 176–177
Phage
  contamination, 5–6
  deletions, 41
  lambda, 22–41
  lambda titration, 28
  P1, 22
  plaque screening by nucleic acid hybridization, 147–149
  purification, 33–35, 40
  R17, 5, 198
  stock growth, 149
  stocks, 30
  T1, 4–5, 31
  T2, 5
  T4, 5
  tails, breaking, 31
Phase partition
  phenol, 99–100
  dextran–polyethylene glycol, 64–68
Phenol
  neutralization, 100
  optical density, 100
  extraction, 99–100
  reagent, 76
Phenyl methyl sulfonyl fluoride, 16
Phosphate buffers, 177–178
Phosphocellulose, 69–73
Phosphorylated small molecules, 52

Photography, 181–186
  fluorescence of DNA, 90–92
  gels, 121–122
Pipetting concentrated sulfuric acid, 48
Plaque hybridization, 146–149
Plasmid-containing colony screening by nucleic acid hybridization, 146–151
Plasmid copy number amplification, 101, 102, 105
Plasmid DNA purification, 97, 101–106, 106–108
Plasmid strain storage, 136
Plastic containers, 173–174
Plastics, some physical properties, 173–174
Plate stocks, 29–30, 149
PMSF, 16
Polaroid film, 122
Polaroid slides, 181
Polyacrylamide gel electrophoresis of
  nucleic acids, 114–124
  proteins, 78–88
Polyethylene glycol, 64–68, 97–98
Polyethylenimine, 73
Polymin P, 73
Polypropylene, 173
Polysomal RNA from *Drosophila* embryos, 166–169
Polystyrene, 173
Pouring columns, 70–71
PPO, 88
Precipitating
  nucleic acids, 61, 62, 64
  ethanol, 95–96
  PEG, 97–98
  TCA, 96–97
Preflashing film, 185
Preparative centrifuges, 190
Proline uptake, 5
Pronase, self-digestion, 111
Prophage, 38
Protamine sulfate precipitation, 73
Protein
  acrylamide gel electrophoresis, 78–88
  ammonium sulfate precipitation, 62–64
  concentrating, 76–77
  concentration, determining, 74–76
  heat precipitation, 107
  lability, 77
  purification on columns, 69–74
  purification, large-scale, 62
  purifications, 61–62
  removing nucleic acids, 64–68
  stability, 61
  stabilizing, 77–78
Protoplasts, 18
Purification of nucleic acids, *see* DNA
Purification of poly-A^+ RNA, 168–170
Purification (*see* DNA isolation)
  proteins, 11, 62–64, 68–73
  phage, 33–35, 40
Purity
  plasmid DNA, 106
  nucleic acid, 89
Pyrophosphate as a chelator, 41

R17 and P1 plates, 198
R17 antiserum, 6
Rabbit reticulocyte lysate preparation, 163–166
Radioactive gels, 78–88, 114–124
Radiolabelling the 5′ ends of RNA or DNA, 143–145
Rare mutants, 35
Rare recombinants, 26
*rec*A, 21–22
Recessive marker transduction, 23–24
Recipe list, 195
Recombinant DNA, 129–172
Recombination, stimulation, 36
Recovering CsCl, 188
Replica plating, 24
Resistance to phage infection, 37
Restriction digestion conditions, 125–127
Restriction-modification, 22
Restriction site adapters, 129
Reticulocyte system, 163–166
Reverse transcriptase, 141
Reversion background in P1 transduction, 23
Ribitol plates, 200
Ribosome removal, 62, 64–66
Ribulokinase assay, 52–56
Ribulose excretion, 49
RNA
  extraction, 166
  filter blots, 152
  gel electrophoresis, 114–124
  hybridization, 145
  hybridization to filter-bound DNA, 156
  hydrolysis by high pH, 143
  polymerase, 45–46, 73
  precipitation with TCA, 96–97
  purification, 164–168, 166–172
  size fractionation by sucrose gradient centrifugation, 170–172
  sucrose gradients, 170–172
  storage, 167
  transfers, 152
RNase, in sucrose, 171
RNase-less strains, 4

S1, 130
S-30, 56–60
S_7, 28
Saturation of ammonium sulfate, 64
Scoring phage plaques for mutants, 36
Screening
  colonies, 149–151
  phage plaques, 147–149
  recombinant DNA clones by nucleic acid hybridization, 146–151
SDS, 20, 79, 99–100
  quenching of fluorescence, 91
  episome curing, 20
  precipitation by K^+, 100
Sealing tubes, 50
Segregation of markers in *E. coli,* 24
Selecting RNA complementary to a DNA, 156–159
Separation from monomers and precursors, 96
Sephadex, 68–73
Serine proteases, 16
Shear breaking, DNA, 129–130
Shigella, 22
Siliconization, 94, 174
Single-colony DNA isolation, 151–152
Single-strand specific nuclease, 130
Size fractionation of DNA, 97–98, 114–124
Slides, making, 181–183
Spectrophotometers, nonlinearity, 11
Sodium deoxycholate, 16
Sodium dodecyl sulfate (*see* SDS), 20, 79, 99
Soft glass, properties, 173
Sonication of cells to open, 15
Southern transfers, 152–156, 184
Spheroplasts, making, 18
Spontaneous episome curing, 20
SSC buffer for DNA, 202
Stab-agar for storing strains, 201
Stabilizing proteins, 77
Stacking gel, 79, 81
Staining gels
  protein, 87
  DNA, 121–122
Sterile technique
  airborne phage, 31
  autoclaving reagents, 57
  cacodylate buffers, 178
  carboys, 11–12
  DEPC, 168
  pipettes, 174–175
  phage contamination, 5–6, 31
  phosphate buffers, 177–178
Stock solutions for media preparations, 195
Storage
  lambda, 35
  bacteria, 3, 201
  DNA, 92–93
  plasmid-containing strains, 136
Strain
  pedigrees, 2
  purity, 1
Streptomycin resistance, 20
Sucrose
  RNase-free, 171
  gradients, 170–172
Sulfuric acid, pipetting, 48
Supercoils, isolation of plasmid, 102
Sus phage, lysogeny, 38
Synthesis of NBPC, 159
Synthesizing radiolabelled RNA, 140–145
Synthesizing radiolabelled cDNA, 141–143

T4
  DNA ligase, 73–130
  plates, 199
  polynucleotide kinase, 144
TB plates, 197
TCA, 46, 96
  precipitation, nucleic acids, 46, 96–97
  proteins, 163
TE buffer for DNA, 201
TEAE–cellulose, 69
Terminal transferase, 133
Tet^s cells, selection, 137
Tetrazolium plates, 199
Tetracycline, concentration in transformation, 136
Thymine-minus cells, 5
Titering, 22
T_m
  nucleic acids, 145
  dependence on length, 146
Toluenization, 44, 47, 54
Toothpicking plaques, 36
Transcription–translation system, 56–60
Transformation of *E. coli,* 134–136
Transformation with λ phage, 27
Translation reaction, reticulocyte system, 165
  wheat germ system, 163
Transposons, 26
Trichloroacetic acid, 45–46, 96–97, 163
Triphosphates
  concentration determination, 112
  λ_{max} and molar absorbance, 112
  preparation, 111
Tris buffers, 177
Two-dimensional gels, 79

Urea gels, 79–85
Useful numbers, 205
UV
  absorbance of nucleic acids, 67, 89–90
  absorbance of proteins, 74
  irradiation, 36
  sensitivity, 21
  sterilization, 6
  transilluminator, 91

Vacuum dialysis, 77
Vacuum pumps, 190
Valine resistance, 18
Valine sensitivity, 5
Vertical gel electrophoresis, 78–84
Virulent lambda, 37
Vitamins, 5

Washing
  chromatography resins, 70
  dialysis tubing, 186
  pipettes, 174–175
Wheat germ extract, 161

Xylene cyanol, 118

YT media and plates, 196

Zubay synthesis, 58–60